Exploratory Programming for the Arts and Humanities

Exploratory Programming for the Arts and Humanities

Second Edition

Nick Montfort

The MIT Press
Cambridge, Massachusetts
London, England

The open access edition of this work was made possible by generous funding from the MIT Libraries. This work is subject to a Creative Commons CC-BY-NC-ND license.

This book was set in ITC Stone Serif Std and ITC Stone Sans Std by New Best-set Typesetters Ltd. Printed and bound in the United States of America.

Library of Congress Cataloging-in-Publication Data

Names: Montfort, Nick, author.
Title: Exploratory programming for the arts and humanities / Nick Montfort.
Description: Second edition. | Cambridge, Massachusetts : The MIT Press, [2021] | Includes
    bibliographical references and index. | Summary: "Exploratory Programming for the Arts
    and Humanities offers a course on programming without prerequisites. It covers both the
    essentials of computing and the main areas in which computing applies to the arts and
    humanities"—Provided by publisher.
Identifiers: LCCN 2019059151 | ISBN 9780262044608 (hardcover)
Subjects: LCSH: Computer programming. | Digital humanities.
Classification: LCC QA76.6 .M664 2021 | DDC 005.1—dc23

LC record available at https://lccn.loc.gov/2019059151

10   9   8   7   6   5   4   3   2   1

[Contents]

[List of Figures]

[Acknowledgments]

First, my thanks go to everyone who has taught and encouraged exploratory programming, from at least the 1960s through today.

Many people discussed the concept of this book with me at different stages of the project. Some of these informal conversations were quite important to the final direction I took with the text—for instance, I learned that senior artists and researchers were interested in learning programming using a book of this sort. In response, I developed a book that could be used in a class or by independent learners and tried to improve both of these aspects of the book in the second edition.

In developing the first edition, I greatly appreciate the opportunity to do daylong and multiday workshops on exploratory programming in New York (at NYU), in Mexico City (at UAM-Cuajimalpa), and in the Boston area (at MIT). I also had the opportunity to teach an undergraduate/graduate course based directly on a late draft of the book in New York, at the New School, thanks to Anne Balsamo. My New School students were a great help to me as I worked to complete the first edition. I've benefited from many experiences teaching programming, and also wish to thank my students in semester-long courses at MIT, the University of Pennsylvania, and the University of Baltimore. My thanks go to those at the New York City gallery Babycastles, where a good bit of work on this book was originally done, and particularly to those who helped out there by reading and commenting on parts of the manuscript as I completed it: Del, Emi, Frank, Justin, Lauren, Lee, Nitzan, Patrick, Stephanie, and Todd. As I was in the last stages of work on the first edition, I was able to teach a two-day course based on some of it for the School for Poetic Computation in New York City. I thank my students and those who ran the school's summer session for this opportunity. I have been fortunate to learn about programming and computing throughout my life in many contexts; an important one was at Penn, where I was particularly helped by Michael Kearns, Mitch Marcus, and the researchers at the Institute for Research in Cognitive Science.

My thanks go to several who reviewed the full text of this book in both of its editions. Erik Stayton went through the full first edition manuscript, commenting on it and correcting it, and also completed all the exercises. Patsy Baudoin provided detailed comments on a full draft of that manuscript. The second edition benefited from the close attention of Judy Heflin, who reviewed the text closely and also served as my teaching assistant when I used a draft of this edition. Todd Anderson did a technical edit of the book to help winnow out many mistakes that remained. My inestimable spouse, Flourish Klink, read and commented on the book and supported me in very many other ways as I worked on this project.

As I developed a draft of the second edition and began to refine it, I got to lead a months-long discussion about the draft book, and an in-person workshop based on it, with members of the Society for Spoken Art, and I greatly appreciate the opportunity to refine the book in this creative context. Thank you all, ITs of Full Circle. Thanks, too, to my students at MIT who were the first to use a full draft of the second edition in a regular classroom course: Andrea, Casey, Janina, JJ, and Meng Fu.

The conventional wisdom about a new edition of a textbook seems to be that it's simply a ploy to milk more money out of students. It would be hard to accuse anyone of that this time around, because the second edition will be available both in a print format (which I believe is extremely well-suited to learning, and worth buying or borrowing from a library) and in a free, digital and screen-based open-access format. That's due to the generosity of the MIT Libraries, which offered the financial support needed to make this edition freely available. Beyond that, the conventional wisdom is quite off the mark, as anyone who has worked on a substantial textbook revision like this one will know! I would like to thank everyone who reported errors in the first edition. I also want to give special thanks to two groups of people whose use of the first edition helped me get to this point. One is a group of faculty and staff at Trinity College in Hartford, CT, who undertook extensive study of the book; thanks to an invitation from Jason B. Jones, I was able to talk with the group and hear presentations about where their studies had led them. The other people who helped immensely were instructors who had used the first edition and could report how it worked (or didn't) in their particular class contexts. I particularly want to thank Angela Chang and Zach Whalen for extensive discussions of their teaching experiences.

The MIT Press of course arranged for anonymous reviewers to consider both the original proposal and the manuscript closely; I am grateful to these reviewers for their support of the project and for their valuable comments and suggestions. At the MIT Press, I also particularly would like to thank Doug Sery, who has discussed, worked on, and supported my book projects for more than a decade and a half now. I can't imagine having explored as many issues in digital media and creative computing, in the same breath and depth, without his backing over the years. I also appreciate the work MIT Press editor Kathy Caruso (who worked on three previous books of mine) did to improve the manuscript and ready the book for publication, in both editions. Any errors in the published text are of course my responsibility.

The text of appendix A, along with a few paragraphs from the introduction, was also published in *A New Companion to Digital Humanities*, edited by Susan Schreibman, Ray Siemens, and John Unsworth, 98–109 (New York: John Wiley & Sons, 2016), as "Exploratory Programming in Digital Humanities Pedagogy and Research."

[1]   Introduction

[1.1]

This is a book about how to think with computation and how to understand computation as part of culture. Programming is introduced as a way to iteratively design both artworks and humanities projects, in a process that enables the programmer to discover, through the process of programming, the ultimate direction that the project will take. My aim is to explain enough about programming to allow a new programmer to explore and inquire with computation, and to help such a person go on to learn more about programming while pursuing projects in the arts and humanities. That is, when someone finishes going through this book, typing in the code as requested and doing the exercises and projects as requested, that person will be a programmer—not particularly an expert, but a person with the ability to use computation in general ways to explore the arts and humanities. The person who completes the work in this book may be a beginning programmer but will be equipped to explore areas of intellectual interest and will be ready to learn more about programming as it becomes necessary.

This book is mainly addressed to people without a programming background— particularly to both individual, self-directed learners and to graduate students in the arts and humanities. I developed this book in part for use in university courses, as a textbook. I also put a lot of effort into developing a book that will be useful outside a standard classroom and course. I provide suggestions for teaching this book, and for learning from it in a class, in appendix B, "Contexts for Learning," which also includes some suggestions for self-directed students.

To some, *programming* is associated with expertise, professional status, and esoteric technical difficulty. I don't think the term *programming* needs to be intimidating, any more than the terms *writing* or *sketching* do. These are simply the conventional words for different activities—creative activities that are also methods for inquiry.

You don't need any background in programming to learn from this book, and courses based on this book do not need to require any such background. If you are already comfortable programming, you may still benefit from using *Exploratory Programming for the Arts and Humanities*, particularly if your work as a programmer has been instrumental: that is, you have mainly worked to implement specifications and solve specific problems rather than using programming to explore. I have assumed, though, that a reader has no previous programming experience.

In my approach to programming, I seek to show its exploratory potential—its facility for sketching, brainstorming, and inquiring about important topics. My discussions, exercises, and explanations of how programming works also include some significant consideration of how computation and programming are culturally situated. This means how programming relates to the manipulation of media of different sorts and, more broadly, to the methods and questions of the arts and humanities.

WARNING! Do not read past this chapter without having your computer next to the book, if you are lucky enough to have the printed version of this book. If you are using the open access digital edition, read it, if at all possible, on a separate device that you place next to your computer.

This book was written and designed to be read alongside a computer, allowing the reader to program while progressing through the book. The book is really meant to be part of a human-book-computer system, one that is set up to help the human learn. After reading this introductory material, I consider it essential to use this book while programming.

I ask at times that readers follow along and simply type code in directly from the book, in part to gain familiarity and comfort with practical aspects of inputting and running programs. At other times, readers are asked to try modifying existing programs, sometimes in minor ways, sometimes in more significant ways. Some carefully chosen exercises, initially simpler and then more substantial, are provided. In some of these cases, a specification is given for a program, describing how it is supposed to work. Although doing such exercises is not an exploratory way to program, they are included to provide a wider range of programming experiences and foster familiarity with code and computing. Particularly when learning the fundamentals of programming, such exercises are important. For these reasons, these exercises are not evenly distributed throughout the book; many of them are provided when fundamentals are introduced and shortly thereafter.

Finally, throughout the book, "free projects" are described. This term is one I made up, meant to express that these aren't the standard sort of exercises or problems. These are intentionally underspecified projects. The idea is that each free project leaves room for learners to determine their own directions and to write different sorts of programs.

You are free to set your own goals and directions, within a general framework. Most of these projects should be done several times, as I indicate. In these projects, the final program is to be arrived at not simply by implementing a fixed specification, but by undertaking some amount of exploration through programming. Because of the unusual way I ask that learners use this book, I go through the four main methods of using the book in detail at the end of this chapter, in 1.8, Programming and Exploring Together.

In some books and courses on programming, readers learn about different sorting algorithms and about how these algorithms have different complexities in space and in time. These are fine topics, and necessary when building a deep foundation for those who will go on to understand the science of computation very thoroughly. If you know already that you are seeking understanding and skills equivalent to a bachelor's degree in computer science or that you actually wish to pursue such a degree, you should probably find a more appropriate book or take a course that covers that material. Furthermore, if your interest is only in learning how to program, and not in the use of programming for exploration or in making a connection between computation and culture, there are good shorter books worth considering. One example I provided in the first edition is Chris Pine's *Learn to Program*, second edition, which uses Ruby to teach general programming principles. Similarly, if you really don't wish to learn to program, but you are interested in how computation relates to issues in the arts and humanities, there are plenty of books and articles focused on these relationships. One of them is a book I wrote with nine coauthors, *10 PRINT CHR$(205.5+RND(1)); : GOTO 10*, which studies a one-line BASIC program in great depth, considering many of its cultural contexts.

It's fair at this point to be clear about what *Exploratory Programming for the Arts and Humanities* is not: It is not an attempt to have readers completely understand any one particular programming language. It is not meant to show people how to professionally produce products or complete the typical "deliverables" of software engineering. It is, as stated earlier, not meant to provide a complete first course for those who will continue to do significant study of computation itself. This book is certainly compatible with computer science education, but it is an attempt to lay a different foundation, a foundation for artistic and humanistic inquiry with computing.

This is a book about how to think with computation—specifically, how to think about questions in the arts and humanities and how to do so by means of programming. I believe this book will supply a solid enough foundation so that those who read it, and who follow along and do the exercises, will be able to pursue a greater ability in specific programming languages, learn about essential ways of collaborating on software projects, and understand that computation significantly engages with culture and with intellectual concerns in the arts and humanities. I also hope that working through this book will reveal the potential of programming as a means of inquiry and art-making—at least, that it will reveal *enough* of that potential. As long as it does, readers will likely be eager to continue their programming, and their inquiry, after they are done with the structure I have offered here.

[1.2] Exploration versus Exploitation

Exploration as an activity and an outlook is discussed in many contexts; let's consider a very everyday way in which people might explore.

On a trip to the grocery store, a shopper may not be interested in locating new produce that has just started to be distributed locally or other new foodstuffs that have been introduced. Instead, this shopper may simply *exploit* existing knowledge about what already known food is appropriately nourishing, inexpensive, and pleasing to the taste. Such a shopper would probably be able to remember where such foods are located within the store, allowing a supply of groceries to be quickly gathered. The result is efficient, and I admit, I myself almost always shop in this way. But we should not expect culinary breakthroughs or novelty in diet from a person who only exploits existing knowledge in this way. A shopper who does not wish to look around even within a store, and certainly not beyond that store to a new farmer's market or specialty shop, will be good at continuing past successes (and failures) and will also have a harder time discovering new options. This will particularly be the case if the shopper is not prompted to think of new food choices because of interesting restaurant experiences or by learning about new foods from others.

In organizational behavior, machine learning, and grocery shopping, it is desirable to balance exploration with exploitation. We can imagine a shopper who does nothing but explore—who tries new foods at random but never returns to enjoy a particularly pleasing food again. What is being learned from such exploration? As described, nothing at all is being learned. Not just very little, but nothing: each random selection is completely independent of the previous ones. One day's grocery basket could be improved by remembering some of the best items found so far and selecting those while also continuing to look for new food. Most grocery-seeking individuals balance exploration and exploitation in some way, just as successful companies try to profit from existing, stable lines of business while they also try out new opportunities that might pay off significantly. A robot finding its way around a changing or partially known environment should exploit some known ways to get from place to place while also devoting some time to exploration, in the hopes that it can find more efficient routes.

It's true that *exploration*, particularly when paired with *exploitation*, has some negative connotations—hinting, for instance, at the history of colonialism. There are some uncomfortable terms and metaphors in computing, as you will know if you have ever read the instructions on how to slave a hard drive to the master. I think some of these terms should be changed, and I find the indifference of some technically oriented individuals to them to be rather disturbing. I don't find that *exploration* packs nearly as much unambiguous historical negativity as *slave*, however. In fact, I am offering it as an alternative to *mastery*, either of the data or of the computer. The point of this book is not to train anyone to be a colonizing power, but to invite people to encounter new ideas and perspectives. The exploration one can do with computing is, to me, more along the lines of how someone from the United States might visit and explore Paris or Mexico City and learn about different architectures, cuisines, organizations of urban spaces, and histories. So perhaps it makes some sense to accuse me of being touristic, but I don't think—whatever the ways in which colonialism is wrapped up with the cities of the world—that my perspective encourages the most negative sense of exploration.

The idea of exploratory programming is not supposed to provide the single solution or "one true way" to approach computing; it's not a suggestion that programmers never develop a system from an existing specification. It's meant, instead, to be one valuable mode in which to think, to encounter computation, and to bring the abilities of the computer to address one's important questions—artistic, cultural, or otherwise.

A problem with programming, as it is typically encountered, is that many people who gain some ability to program hardly learn to explore at all. There are substantial challenges involved in learning how to program and in learning how computing works. If one is interested in fully understanding basic data structures and gaining the type of knowledge that a computer science student needs, it can be hard to discover at the same time, at an early point in one's programming experience, how to use programming as a means of inquiry.

Linked lists and binary trees are essential concepts for those learning the science of computation, but a great deal of exploration can be done without understanding these concepts. Those working in artistic and humanistic areas can apply programming fundamentals to discover how exploration can be done through programming. They can learn how computing allows for abstraction and generalized calculations. They can gain comfort with programming, learn to program effectively, and see how to use programming as a means of inquiry. For those who don't plan on getting a degree in computer science, it can sometimes be difficult to understand the bigger picture, hard to discern how to usefully work with data and how to gain comfort with programming, while dealing with the more advanced topics that are covered in introductory programming courses. That is, it can be hard to see the forest for the binary trees.

Beyond that, many of those who haven't yet learned about programming get the impression that it is simply a power tool for completing an edifice or a vehicle used to commute from one point to another. While the computer can have impressive results when used in such instrumental ways, it can be used for even more impressive purposes when understood as a sketchpad, sandbox, prototyping kit, telescope, and microscope. As a system for exploration and inquiry, there really isn't anything else with the same capabilities as the computer. Exploratory programming is about using computation in this way, as an artist and humanist.

[1.3]   A Justification for Learning to Program?

As I wrote this book, I looked at the beginning of several popular books on learning to play guitar, on learning Spanish, and on introducing artificial intelligence. Interestingly, none of these books seemed to have a section justifying in detail why one should learn to play guitar, or learn Spanish, or understand the basics of artificial intelligence. Sometimes there were a few sentences about how cool it is to play the guitar, or some facts about how many people in the world speak Spanish, or some information about how interesting AI is as a field, but there was nothing like a pep talk, rallying cry, or other exhortation to learn the subject matter of the book. At most, these books just say something like "So you want to learn to play guitar?" or "So you're ready to learn Spanish?"

So—you want to learn to program as a means of creativity and inquiry?

I have found many people interested in learning how to program as a way of exploring computation, texts, language more generally, images, sound, historical and spatial data, and other things. Not everyone feels this way, but when you start looking for such people, you find that a surprising number do. If you are one of these people, this book is for you.

For those who actively fear computers and mathematics, I suggest resolving to overcome this feeling before getting started as a programmer. People who fear reading and conversation and do not believe they are able to learn a foreign language are really not well-suited to begin studying Spanish; they should somehow gain at least a willingness to read and talk before they begin a course of study. People who have convinced themselves that they are inherently nonmusical are probably not in the best state of mind to begin learning guitar or piano. They should first be willing to try to make music.

Not everyone is convinced that programming is an important way of intellectually exploring in the arts and humanities—even within the area known as the digital humanities, for instance. Some people see programming as a technical detail, something peripheral that can be contracted out or outsourced. Or perhaps there's always an app for it, so that whatever project one has in mind, it can be done without programming. In this introduction, I discuss some of the roles that exploratory programming has played in culture and computing, as I work to motivate this particular practice and approach, but in this part of the book I do not make a full, extensive argument for programming being a core experience and ability of those in humanities and the arts. If you meet someone who is unclear on why artists and humanists would program, and you would like to share some other perspectives on the importance of programming, I offer appendix A, "Why Program?," to supplement your discussion.

[1.4]   Creative Computing and Programming as Inquiry

My aim in this book is to help new programmers see the creative potential of the computer and understand how computation can be used to explore and inquire. My plan is to provide some insight into how this can be done, to help readers reach a certain level of comfort with programming, and to show how new programmers can pull themselves up using these bootstraps and can continue to learn in whatever specific domains are interesting and compelling.

This book covers ways to develop digital art of various sorts, literary, visual, and sonic, but one does not have to have a practice as a poet, artist, or musician to work through these examples. The book also shows how one can undertake several types of analysis and inquiry with textual and image data and with data of other sorts. To benefit from working through such examples, though, a programmer does not have to be strongly focused on the particular media being examined or the specific type of humanistic perspective taken.

In very general terms, this book covers two types of programming practice. One is programming to create new types of artworks, very broadly speaking—works in both revered and popular forms, literary and sound work, and more. This includes creating games, bots, and many other computer programs that don't have clear analogues in the fine arts. I call this area *creative computing*, a term that I directly ripped off from an influential magazine, *Creative Computing*, that was published from 1974 to 1985. To me, this is a pleasing and appropriately inclusive term. The term also serves as a reminder that society was onto something in the late 1970s and early 1980s, when popular access to computers and programming was increasing and there was a great deal of enthusiasm for the cultural potential of the computer.

The other practice is one I call *programming as inquiry*. This is the use of programming to probe data and come up with new ideas. One doesn't have to have a large dataset to use programming to inquire. A program without much data can help one understand how randomness works or offer some understanding of properties of the Fibonacci sequence. In analyzing corpora of photos, an exploratory researcher might learn about significant but unnoticed differences in how frequent nighttime and daytime photography is and whether those who post photos prefer indoor or outdoor photography. These sorts of higher-level differences might be noticed after starting with an analysis of lower-level features, such as brightness and contrast. I think of this use of programming, programming as inquiry, as helping scholars and researchers explore ideas and determine what more complex projects should be undertaken.

Now, the truth is that creative computing and programming as inquiry are not actually two distinct categories. A program developed for humanistic inquiry about data is itself a creative production, a digital artifact representing a new way of thinking. A program to generate texts or to show a visual animation is also an exploration, an inquiry, into how words fit together and how shape, line, and color work with one another in motion. You can see examples of this in the work of poet Charles O. Hartman, who helped develop the system PROSE for analysis of poems, including metrical analysis, and who used similar techniques and ideas to computationally generate poems, as seen in his and Hugh Kenner's book *Sentences*. His work in programming as inquiry (to analyze poems) and in creative computing (to generate poems) were two sides of the same coin (Hartman 1996).

Really, then, my two terms are different names for the same type of activity. The two terms don't describe distinct practices, but rather two stances, two different types of emphasis within the same area of practice. That's why I think it makes perfect sense to learn about exploratory programming for *both* the arts and the humanities.

[1.5]   Programming Breakthroughs

Here's a selective but worthwhile review of the history of computing, focusing on how some of the most important breakthroughs *weren't* accomplished. Google's search engine was *not* fully planned out by expert nonprogrammers and then implemented later; Sergey Brin and Larry Page coded it up themselves as they invented it. Tim

Berners-Lee *didn't* develop the World Wide Web by figuring out all the concepts, applying for a grant, and then dealing with things like programming later. Joseph Weizenbaum *didn't* just design the first chatbot, Eliza, and then get someone else to do the programming later. Douglas Engelbart *didn't* plan all the major advances of his famous system, called oNLine System (NLS)—hypertext, videoconferencing, the mouse, and many others—and then deal with the programming afterwards. Grace Murray Hopper *didn't* come up with the idea of the compiler, and high-level programming languages, without understanding programming to begin with.

These major advances, not simply technical advances but also ones that were resonant with many cultural implications, were *not* made by treating programming as instrumental or as a detail to be handled later. These and dozens of other major breakthroughs were made by programmers who used programming to think, doing creative computing and programming as inquiry. These innovators actually developed their breakthrough ideas while programming.

Douglas Rushkoff, in his book *Program or Be Programmed*, notes that while "we see actual coding as some boring chore, a working-class skill like bricklaying, which may as well be outsourced to some poor nation while our kids play and even design video games . . . the programming—the code itself—is the place from which the most significant innovations emerge" (Rushkoff 2010, 137). The examples I provided are only a few of the historically significant cases in which innovation was inseparable from programming.

The problem with separating high-level digital media design from programming is that, in many cases, they are inseparable. Even if you do plan to enlist others to do the heavy lifting of coding, you will somehow have to figure out what to do in the first place, which requires an understanding of computation, knowing about the capabilities of the computer in the way that programmers do. An article in *Mother Jones* explained this:

> The happy truth is, if you get the fundamentals about how computers think, and how humans can talk to them in a language the machines understand, you can imagine a project that a computer could do, and discuss it in a way that will make sense to an actual programmer. Because as programmers will tell you, the building part is often not the hardest part: It's figuring out what to build. "Unless you can think about the ways computers can solve problems, you can't even know how to ask the questions that need to be answered," says Annette Vee, a University of Pittsburgh professor who studies the spread of computer science literacy. (Raja 2014)

The book you are reading seeks to enable new programmers to do the type of sketching, exploration, and iterative development that were done in the influential projects just listed—and it focuses on humanistic and artistic inquiry. It's my hope that it will help the reader acquire some of the sense for programming, and some of the willingness to explore, that was exhibited by the people who made these important breakthroughs.

[1.6] Programming Languages Used in This Book

Python and Processing are the main languages used in this book, although the first encounter with programming is via JavaScript—in case you thought that all programming languages had to begin with the letter p. In chapter 16, "Sound, Bytes, and Bits," we'll write not entire programs, but the arithmetic expressions (canonically used in the C programming language) that are central to a curious and compelling type of sound generation.

Python is a powerful, standardized, widely used language; I find that it is very good for exploration and also very suitable for new programmers. With additional modules, it can be used for image processing, to develop games, to do extensive statistical work, and for all types of purposes. Without installing anything additional, it can serve very well for simple text processing and to introduce computing. Guido van Rossum began developing Python in 1989. Python is actually included in OS X and GNU/Linux distributions, but I ask that everyone using this book install the same standard distribution of Python 3, Anaconda, which works on OS X, GNU/Linux, and Windows and provides important and very useful features.

Processing is a language, based on and similar to Java in many ways, that provides excellent facilities for computational visual art and design. It was created by Ben Fry and Casey Reas and first released in 2001. Processing (we will use version 3) includes an elegant, simple integrated development environment (IDE) that makes sketching with code easy, and was designed with artistic exploration in mind. For those interested in continuing to explore using Processing, there are several good books and extensive online resources, which appear in the references and are listed, with discussion, at the end of chapter 9. Processing is available at no cost for GNU/Linux, Mac, and Windows.

The language commonly known as JavaScript has a very obvious virtue: it can be incorporated into HTML and can run in practically any Web browser, locally or over the Web. JavaScript as it existed for many years, in the context of HTML and CSS on the Web, may be thought of as something of an affliction, attested to by the fact that *JavaScript: The Good Parts* is a 172-page book, while *JavaScript: The Complete Reference, Third Edition* weighs in at 976 pages. (As an exercise, I invite you to compute the percentage of JavaScript that is not good.) As recently as a few years ago, producing

JavaScript code that worked consistently across browsers required time, effort, and expertise. Even then, one could quickly understand that there are benefits to using this language. Understanding and modifying some existing JavaScript programs would show that they can be very easily shared online. The language, which was released in 1995, is suitable for artistic explorations and for sharing creative projects with others on the Web. It was developed (originally by Brendan Eich) for use by those who weren't computer scientists or professional programmers. Only a text editor and a Web browser are needed to write and run JavaScript, although the Firefox browser offers a scratch-pad for writing, running, and editing code along with other facilities. There are online options for doing similar things, too.

I have to note that when the first edition of this book was published, my feelings about JavaScript were rather negative—and I think justified! Since then, I have done more work in a more recent version of JavaScript, officially called ECMAScript 6 (ES6). This version introduces some pleasing improvements; also, current browsers are providing much more consistent support for ES6 than has been the case in previous years and with earlier versions of the language. This means that while a JavaScript programmer's time used to be consumed in dealing with special cases for different browsers, it's now possible—assuming you are programming for current-generation browsers—to spend more of one's time actually dealing with the core computational issues, exploring and creating. More concretely, it's great to see some of the particular constructs I find intuitive in Python, such as looping over every element of a list, implemented in an intuitive way in ES6.

[1.7]   Free Software and No-Cost Software

In describing what you will be asked to download and set up, I have mentioned that any software needed to pursue *Exploratory Programming for the Arts and Humanities* and not already included with your system is available at no cost. I'll add to that now: everything required to follow along in this book is *also* free software. The distinction is not obvious, but it's an important one, particularly for those concerned with the cultural implications of computing and how computing can be used for inquiry and creative work.

Some software that doesn't cost anything, also called *freeware*, can be downloaded and used without a financial transaction. Freeware can still be encumbered in various ways, however. The license that allows use of the software may say that it can only be used for noncommercial purposes, for instance, or only by students, or only by people who have signed a loyalty oath, or only by men. Furthermore, you may be given the software only in executable form, without source code, making it impossible for you or anyone else to fix bugs in it, to expand it, or to adapt it to different needs, including new computers and operating systems.

Because free software comes with access to source code, some people use *open source* as a similar term, sometimes as a synonym. I prefer the term *free software*. The openness of the source code is not the only freedom in free software. There is also, for instance, the freedom for anyone to be able to use the software for any purpose, even if they are not students or have not signed a loyalty oath. In other words, you can be given "open" source code and still have restrictions placed on how you use it. If you believe that computation is a way of thinking and that code should be treated in the best ways that we treat ideas, your real concern is not only with code being open, but with software freedom.

The term *free software* was first used in its current sense in 1985 by Richard Stallman, who had, shortly before, founded the GNU project to develop an operating system that was to be available as free software. (The *G* in *GNU* is pronounced when speaking of the project, although that's not the case when naming the animal.) In March 1985, Stallman's "GNU Manifesto" was published, and he founded the Free Software Foundation (FSF). Of course, many people desired software liberty before 1985, and they acted to promote it in various ways. And people after 1985 have sometimes wanted the software they create and use to be free but haven't explicitly used the term *free software*. When people write very small programs or snippets of code, for instance, they often don't include a lengthy license or an official declaration, even if they wish their work to be freely available and freely shared. (I don't add licenses to very tiny programs that I write, even though I'm glad for them to be shared, modified, and reused in any way.) Stallman, in 1985, brought together several useful principles to form the first concept of free software, one that has continued to evolve and to be refined—although the principles remain the same.

The FSF's definition of *software freedom* at gnu.org/philosophy/free-sw.html.en includes four points, numbered 0 to 3. (A moment of foreshadowing: As we begin working with lists and arrays, we'll see that it is conventional in computing to begin numbering a sequence with 0. There are reasons that this convention was established and persists, too, which will be covered later.) The four freedoms are as follows:

[**Freedom 0**] The freedom to run the program as you wish, for any purpose.

[**Freedom 1**] The freedom to study how the program works, and change it so it does your computing as you wish. Access to the source code is a precondition for this.

[**Freedom 2**] The freedom to redistribute copies so you can help others.

[**Freedom** 3] The freedom to distribute copies of your modified versions to others. By doing this you can give the whole community a chance to benefit from your changes. Access to the source code is a precondition for this.

A version of, for instance, some commercial illustration software that is licensed for educational use only, by full-time students, does not offer any of these freedoms. When the user of this software finishes school, they are not legally allowed to use the software to further edit or export illustrations done while in school—denying the student access to their own creative work. A supposedly "free" app from Apple, Inc.'s App Store, even if it allows use for any purpose, still lacks freedoms 1, 2, and 3. A user cannot redistribute copies to a neighbor, family member, or even to himself or herself; it's a requirement that one go to the App Store. If the app disappears from there and the user gets a new phone, too bad. Also, freedoms 1 and 3 are missing because they rely on access to source code. For these reasons, it doesn't make a great deal of sense to me to call such purportedly free apps "free." I think of them as currently priced at zero dollars and as locked down and restricted.

You may hear that the *free* in *free software* is free as in *free speech*, not as in *free beer*. That can be helpful to understanding the concept. People also refer to free software as free/libre/open source, or FLOSS, to emphasize that the relevant sense of free is "libre" as in freedom, not "gratis" as in given away without cost. Another good way to put it is that freedoms in *software freedom* are not really the freedoms of the software itself, but the freedoms of the people using the software; they are user and programmer freedoms (Hill 2011).

Many good speeches have been given, and many articles and books have been written, about the concept of free software and the virtues of this idea. I will offer just a brief comment here. When people innovate and develop new ways of using computation, this can be treated as a contribution to the world of ideas—$E = mc^2$ or the polio vaccine—or it can be treated like the song "Happy Birthday to You," the copyright to which was claimed by Warner/Chappell Music Inc. more than one hundred years after the song was written. If people view the computer as a way to make money fast, to reallocate resources from other people to themselves, they will prefer the latter option. If they believe that computation's most significant use is in enlarging the human intellect and making the world a better place, it will be more reasonable to establish a framework of sharing, freedom to use software for any purpose, and allowing people to build on one another's successes. No-cost software in general, those programs that sport price tags of zero dollars, do not inherently embody this idea. This is, however, the basic concept of free software.

For more about the free software movement, the interested reader can consult the GNU site at gnu.org/philosophy/free-software-intro.html. For an extended history focused on GNU and the development of the Linux kernel, there's the book *For Fun and Profit: A History of the Free and Open Source Software Revolution* (Tozzi 2017). I certainly advocate for free software and encourage people to read about, and join, the movement. If you are going to start programming using this book, however, I suggest getting started soon and returning to learn about free software further once you explore as a programmer for a while. It will mean more to you.

You can follow along in this book using a system of the sort I used to write it—using entirely free software. That is, you can use a free operating system (such as a distribution of GNU/Linux; I run Ubuntu, which is based on Debian) and only use programming environments and a text editor that are free software applications. It is also possible to follow along in this book using the major proprietary operating systems, Mac OS X and Windows. Even the most domineering corporate interests don't yet prohibit people from programming on desktop and notebook computers or from sharing their programs with others online. It is possible to write free software using proprietary software, and, personally, I wouldn't (and don't) reject free software that is written in this way. If one believes in the free software concept, though, it is ideal to use free software throughout one's environment.

Corporate interests do prevent the general-purpose use of many tablets, mobile phones, and videogame consoles. These computers, by design, cannot compute in the most general sense. They are made to deliver approved products. They effectively prohibit users from writing programs on them and then giving those programs away. Instead, all programs must be sold through a corporate channel and (even if they are sold for zero dollars) must be approved by corporate authorities. Programs that don't fit into the corporate framework and don't match the corporate idea of a typical product may not be approved. This is of particular consequence for the noncommercial and noncorporate production—humanistic, artistic, poetic, and related sorts of production—likely to be undertaken by readers of this book.

Perhaps it's all right for computing to be made safe for the mall, to be regularized as commerce. Once this happens, then, we should expect that the art we can access will be limited to what we can get at the mall: prints from the poster store and factory-made canvases touched by the Painter of Light. The same can be said for literature and the humanities: if commerce takes over completely, we can have whatever we want that is humanistic and literary, as long as it too is inoffensive, the sort of thing that can be found in the mall.

Again, this topic is worth returning to after some of the power of programming has been demonstrated. A person who doesn't program, after all, is not likely to miss much when using a locked-down computer system that prevents programming. But the matter of free software versus proprietary software is hardly a technical issue disassociated from culture. It's an essential issue for society, and certainly for the arts and humanities, and for all of those who believe computers can be machines for new types of thought and discovery.

[1.8]   Programming and Exploring Together

First, it's better for *almost everyone* to start exploring programming in a group. This could be a formal classroom, a sort of "book club" of people with similar interests in programming, a group drawn from an existing online community that meets online, or just a pair of people. A lengthy argument in favor of social, communal learning can be found in Yasmin B. Kafai and Quinn Burke's book *Connected Code: Why Children Need to Learn Programming*. Although the argument there is about the way children learn programming, it's quite applicable to people in general.

When you work in such a group, you have someone to ask for help, people who can see your creative work (your free projects), and other people's projects to admire and learn from. My suggestion is that you, as a new programmer, do individual work on exercises and projects and then collaborate in discussion, critique, and appreciation. If you like, you can also collaboratively rework and remix these projects you undertook individually. It can be great to work in a group on particular projects, too, but there is some danger that new programmers may leave the coding to whoever seems more expert at the time, and this will hinder other people's learning—and prevent anyone from learning good methods of collaboration.

Now, I've stated that learning in a group is better for *almost everyone*. Learning in this way provides generally positive social pressure to progress through the book and offers a way to see others' response to one's free projects. It places programming in the social world, which is where it indeed actually exists for those who program. Some people prefer not to have this social pressure and this response from others, however.

I have to consider that people *do* become better writers and improve the way they think about their lives by writing diaries, which are written individually and are not intended for anyone else to read. "There is delight in singing, tho' none hear / Beside the singer," as Walter Savage Landor wrote. So although a social context for programming education can be very important, if working in a group as you first learn to program seems intimidating to you, you can still learn to program.

Whether you are working in a group or not, I must emphasize again that you should definitely work through this book—all of the rest of the book—*with a computer in front of you*. And it should be a computer you are actively using! Typing code from the book into your programming environment will give you a feel for typing in valid code and running it and will reassure you that short programs can be entered from scratch and run. As you work through this book, I suggest you avoid setting aside some fixed, sacrosanct time that is the *only* time when you program or (even worse) going to some sterile computer lab that you would never otherwise visit and making this the exclusive place to do your work. Just set up the systems you need on the computer that you regularly use and make time the way you would for other types of study, practice, and activity.

You don't need to be online as you work through this book, except to download software. The Processing reference pages, for instance, are part of the Processing download—you can access them through the Help > Reference menu item. You also *can* access them online, but you have them installed on your system, and you can be sure that the ones you have installed are the right version. In the Anaconda Navigator, you can click on the Learning tab on the left and you'll see plenty of Python documentation, including Python Reference. This happens to link out to the Web and seems to always point the latest version of the Python 3 documentation, even if you have an earlier version of Python 3 installed. This book shows you how to get basic information about how Python works within the interpreter we will use, Jupyter Notebook. The point is, it's not necessary to be online to learn more about the specifics of Python or Processing—certainly not to do the exercises. Reading through such reference material is not even essential for exploration and for doing the free projects.

Thus, if you find it distracting to be online, you might take your notebook computer to one of the dwindling places in the world where you won't be online. If you find yourself on a plane or train without Wi-Fi, or otherwise in an environment in which you lack Internet access, don't let that stop you from working through *Exploratory Programming for the Arts and Humanities*.

However, if it isn't a distraction and net access is available, there are some good reasons to be online. In addition to the Processing reference pages you have on disk and your ability to find out how Python works, you can also check online documentation. Once you understand the essential concepts, the documentation for Python and Processing can help you determine how to accomplish what you conceptually understand in the specific syntax of the programming language. For Python 3, documentation for the current version can be installed locally and is also available online at docs.python.org/3/. The Processing reference, again, gets installed locally with Processing itself, but also be found online at processing.org/reference/.

Beyond this, you can check existing forum posts about these languages, and of course ask questions online yourself, either on forums or in real time. Your real-time interlocutors might be people in your programming group who are part of a video conference with you or in a chat room. Even if you have a local group, it can be nice to be online some of the time, as you can copy and paste code you have written along with questions you have.

While I always encourage exploration, piecing together snippets of code you find online—if you don't comprehend them—will hinder your learning rather than helping you. Remember that your main goal here is to understand programming as a method of inquiry and creativity, not simply to hook together a working (but incomprehensible) code contraption. Yes, it's fine to read what other sources have to say about what Python and Processing offer and to look at programming examples beyond this book. Keep in mind, though, that what I introduce in the book is meant to be adequate for the exercises and free projects. You do not need to troll online for some function I am keeping secret. And should you locate code elsewhere that you incorporate in your projects, you need to be sure that you understand everything about how it works.

With very few exceptions, programming in the proverbial "real world" is a social activity. Back in the day, some game programmers at Atari may have secluded themselves for months as they developed cartridges for the VCS (aka the Atari 2600) that were essentially one-person projects. Even then, these programmers did share code and ideas. A student working on a dissertation that involves programming might write the associated code more or less by themselves, but in high-level research meetings and by consulting with others (online and offline) about lower-level programming issues, even such an individual project will typically benefit from discussion with others. Those are some of the most extreme cases of "solo" programming, and even those involve collaboration. Many academic and creative projects involving programming proceed collaboratively, and practically all companies incorporate collaborative programming practices, such as pair programming and code reviews. While we aren't doing industrial development, it's interesting to note an estimate that in industry, "developers spend 30 percent of their time working alone, 50 percent working with one other person, and 20 percent working with two or more people" (Kafai and Burke 2014, 52).

[1.9]   Program as You Go, Testing Yourself

Starting in chapter 3, you will be asked now and then to make a guess about what will happen at various times. When I ask you to do this, I am not suggesting any classroom assignments for grading. These are opportunities for personal reflection. They allow readers to commit to a particular perspective (or at least to a guess). If you follow through with this advice and commit to a guess (writing your guess down or typing it on a virtual sticky note would not be a bad idea), you give yourself the opportunity to see where your thinking is particularly productive and what is left to be learned.

Instructions on how to set up Python, Processing, and everything else that is used in this book appear in chapter 2, "Installation and Setup." I very strongly suggest that you get everything set up before you start. And if you're *teaching* a class in a computer-equipped classroom, I strongly suggest that you see that everything is installed on these computers at the beginning of the semester—while also very strongly encouraging students to install the required free software on their own machines.

It bears emphasizing: to learn how to program a computer, it is important to actually do programming. Programming is a practice, like public speaking, playing a musical instrument, dancing, writing essays or poems, or rock climbing. There's only so much that can be done by reading and thinking about it; to really make any sort of progress, you have to actually undertake the practice and try out the activity. Fortunately, to program, you don't have to travel to a special site with climbing routes or find a music practice room. You just start typing. So if you're actually interested in becoming a new programmer, rather than just admiring my prose style, please install the software required and use it.

You may find it more valuable, particularly as you start to program in an exploratory way, if you program frequently (every day, perhaps) for a short time each session rather than having only one longer time each week set aside. Programming is not a physically strenuous activity, and while it requires some special types of thinking, the kind we are doing does not require the intense concentration and specialized environment of, say, surgery. It doesn't need to be treated as an extraordinary, ritual activity. By the time you near the end of the book, you may wish to devote longer periods of time and more concentration to working on complex projects. But treating programming as overly special, and as requiring a massive block of time, could end up giving new programmers more excuses to avoid programming. All of that said, programming does need to be undertaken often and extensively—it is a practice that needs to be practiced. With that explained, I will describe the ways I believe it is essential for all learners to make use of this book.

[1.10]   Four Methods to Best Learn with This Book

There are four methods of learning with this book, which, when used together, will allow you to learn to program, whether you are in a short workshop, a semester-long class, a small study group, or using the book individually.

[Method 1]   Read and Study the Text

This needs no elaboration. It's just important to emphasize that it is *only* one of four things you need to do with the book!

[Method 2]   Type in the Code as You Read, Run It

By following along through the chapters and typing in code, you will get a *feel* for programming. You will understand some of what's involved in writing code, locating errors, and having an expectation about what will result before you try it out. Hopefully this will give you an idea of what the *practice* of programming is like, in the sense of a clinical practice, an artistic practice, or perhaps even a spiritual practice.

If you were to just read the book without typing in the code I've provided, you will not get a feel for programming. You will just get a feel for reading a book. You know how to read books already, so this won't be useful.

If you are lucky enough to use this book in a classroom context, you will have another way to get a feel for programming. The instructor can model for you how programmers sketch and inquire. You can see someone, the instructor, trying out ideas by programming right in front of you and doing some different things beyond what the book offers. But whether or not you have that benefit, I want to provide within the book some idea of what programming is like, at a specific and low level. To do that, I ask you to type in code as you follow along with my written discussion of programming.

I think it's very important to get a feel for programming. By itself, however, typing in code can be boring, because when I ask you to follow explanations and type things in, you are not doing much independent thinking. You're not testing yourself to see if you have learned the fundamentals of programming or of manipulating different media using code or of analyzing different sorts of data. So a feel for programming is not enough by itself, and this is only one of four ways this book can help you learn.

[Method 3]   Do the Exercises

There are several more traditional "exercises"—essentially, problems for you to solve—that are presented in the book, mostly early on when programming fundamentals are introduced. You should complete all of these before continuing in the book, because a grasp of the true programming fundamentals is essential. The exercises I have provided are focused on aspects of programming that everyone who seeks to use computation needs to know about.

The exercises can be more frustrating than just typing code in because you may not know what approach to take to solve a particular problem. In these cases, you should review the previous sections of the book and work to understand what is explained there and how analogous problems are solved. The exercises go beyond the previous discussion in some ways, but they are not meant to be tremendous and difficult challenges. They are simply opportunities to do a bit of focused thinking so that, in addition to gaining a *feel* for programming, you can gain *strength*.

The exercises have their limitations, because they ask "closed" questions. There's a particular problem posed, and you are supposed to solve it. Once you have done that, that's it. Fortunately, typing things in and doing the exercises are not the only ways to use this book to learn.

[Method 4]   Define, Pursue, and Share Your Free Projects

Programming exercises can be found in many books, but in a typical first course in programming, even if it is for "nonmajors," it is less typical to let learners define their own projects. I have introduced free projects as early as I think is possible. I have also tried to offer, in these projects, frameworks that connect to what we have studied, along with an openness to the interests of different learners. I believe these free projects are essential to learning about programming from an arts and humanities perspective.

As I've indicated, almost all of the free projects should be done several times by readers of this book. In a workshop- or seminar-style class, students can select the most interesting ones to present.

The three main things you can learn from a free project are how to do something using your programming skills that is *tractable, computational,* and *interesting.* A project has to be tractable—that is, manageable, something you can handle—or it won't be possible for you to accomplish it. Most of the free projects have some basic structure to them but allow latitude in how they are done.

This discussion will hopefully make even more sense after you have learned the fundamentals a good bit using Methods 1, 2, and 3. But to be concrete about the tractable, computational, and interesting at this point, I'll imagine a final project (or "very free project") for a semester-long class, one that is not constrained and builds on everything learned throughout the semester.

For such a project, you may want to develop a networked multimedia system that can converse about art history as well as an expert human can. Nice idea, but this is not a tractable project. Just making a website with some images of visual art and some writing about visual art could be a fine project for a different context, but it isn't a computational project that uses your growing ability as a programmer. Figuring out some simple statistics of different photographs, represented as JPEG files, may be both tractable and computational, but because it isn't very likely to offer insights, it doesn't seem very interesting. It is a challenge to find a free project that meets all three of these criteria. If you do develop a project that interests you, is computational, and can be accomplished, you not only will be developing as a programmer; you'll also be learning about how programming and computation is meaningful to you.

To take this example, you might decide that what interests you is the connection between writing and visual images. If you were doing an analytical (humanistic) project, you might try to use some of the text analysis techniques covered in the book and some of the image-analysis techniques covered in the book, and see, from there, if you can automatically create text/image juxtapositions that are more interesting than random ones. You wouldn't have to end up illustrating a text or writing the caption for an image in order to match words and images in an interesting way. If you want to take a generative (artistic) approach, you could start with images and generate texts based on what you can learn about those images, which might have more to do with global qualities like brightness and color than with what is represented in the images. Or, of course, you could do the reverse: start with texts and generate images based on them. Beyond giving you a *feeling* for what it's like to program and helping you build *strength* as a programmer, these free projects can reveal *why programming matters* to you in particular—why it's interesting from your perspective. They can also give you experience figuring out what's tractable and what really makes use of computation.

For a final project of this sort, you should not only do some sort of a final presentation, but also find out along the way if your work is interesting to you and if it resonates with others. Sharing work in class sessions is an ideal way of doing this. Independent learners face the most difficult challenge when it comes to sharing these projects and seeking meaningful feedback. Teaming up with others or pairing up with another person studying this book, even informally, can be a great help in that regard. While sharing your code and running each other's projects can be ideal, it's not the only way to get useful responses and meaningful critique from others. You can share outputs from your system and raise questions about how it is working via videoconference, by email, on a forum, or on social media.

There are plenty of concepts in each of the chapters that are good and wholesome, plenty that are useful, and plenty that are interesting. Beyond these, there are some that I consider essential for making progress as an explorer of computation—that is, as a programmer who creates and inquires. So while I think it important to know the difference between free software and no-cost software to be an informed citizen, on these lists of essentials I am only including those concepts that one really needs to understand to move on in the book and to become an exploratory programmer, inquiring and creating. If you feel shaky on any of these, it is important to review them and, to whatever extent it's possible, to discuss them with others. You should confirm your understanding of all of these and be able to express that understanding—in speech and, in most cases beyond this introduction, in code.

[1.11]   Essential Concepts

[Concept 1-1]   Explore Creative Computing and Inquiry

This describes the motivation of this book and the direction of any learner using it. Be able to describe what types of programming would be more and what would be less exploratory. Relate exploratory programming to other programming. Characterize creative computing and programming as inquiry, giving examples of recent computer programs you know about.

[Concept 1-2]   Programming Is a Practice and Requires Practice

Understand that learning to program shares important qualities with learning to participate in a sport, play a musical instrument, or speak a new language. How regularly do you plan to practice as a programmer? Where will you program, and will you limit yourself to only special times and places, which may become inconvenient? How have you made your programming environment ready for your use? If the computer you regularly use somehow breaks, will you abandon your attempts to learn programming or find an alternative way to program? Will you give yourself opportunities to program or excuses to avoid doing so?

[Concept 1-3]   Read, Type in Code, Do the Exercises and Free Projects

This book offers four interlocking ways to learn about programming as it pertains to the arts and humanities. Whatever context you are in as a learner, you need to avail yourself of all four ways—reading the book, typing in the code to see what happens, getting through all of the exercises with understanding, and framing and completing the free projects—to develop yourself as a programmer and as someone capable of thinking artistically and humanistically with computation.

[2]   Installation and Setup

[2.1]

We'll use a few different programming languages and environments, which means you will need to install some software. It's important to get things set up now:

- To get an initial sense of computer programs, we'll work very briefly in JavaScript, and for this and some future work, a text editor is necessary.
- The core work in the next several chapters will be done in Python, specifically Python 3, using Jupyter Notebook. We will install a particular distribution of Python 3 called Anaconda. We will use one library that is not included with the Anaconda distribution of Python by default: TextBlob, for text processing. We'll need to install that one as well, which will take a few extra steps.
- In some of the later chapters, we'll work in Processing—specifically, Processing 3.

[2.2]   Install a Text Editor

A *text editor* is a computer program that works on plain, old, classic text files. These files were in the past typically encoded in American Standard Code for Information Interchange (ASCII) and these days are more often in Unicode, which begins with the same characters but has much more representational power and can better encode written symbols from around the world. Not only computer programs (scripts, source code) but also system logs, error reports, README files, and many other sorts of files on a typical computer are text files. They are basically just sequences of characters. This simple format is adequate for many purposes.

A word processor, developed for composing memoranda, reports, papers, letters, and so forth, and capable of various typographical effects, is *not* the same as a text editor. Word processors are often actually very poor at editing plain text files, particularly if those files are computer programs. By default, they will save data in their own native format, not as plain text. They also use proportional fonts by default, which are not desirable for programming. They are set up with models of the English language and will decorate code with squiggly green and red lines to indicate that code is not grammatically proper or correctly spelled according to a standard English dictionary. While this could point out a spelling error in a string or comment somewhere, it's generally not very helpful when programming. You *can* edit a computer program in a word processor, just as you can also drive by turning a car's steering wheel with one of your feet. It's really more effective, however, to use a text editor.

There are many good free software text editors for GNU/Linux, Mac, and Windows, along with other well-functioning ones that are freeware or are sold commercially. I use Atom, which is free software and available on all of these three platforms. (When writing the first edition of this book, I was using Geany, also free software, also cross-platform.) Feel free to help yourself to Atom (or Geany, for that matter). Some people are pleased to use venerable text editors that originated before today's GUI and mouse—GNU Emacs and vi being the most famous. Others might value a text editor that is made specifically for their operating system and follows its interface conventions closely. On the Mac, Brackets is a free software program of this sort. Windows has a built-in text editor, Notepad, which is quite minimal. Notepad++ is a more capable free software editor for Windows and better to use.

Ironically, the simple word processor application TextEdit, which is included with Mac OS X and sure sounds like it should be a text editor, is *not* set up to edit plain text files and actually has to be reconfigured to do so. Download a free software or no-cost text editor, such as Brackets or Atom, instead.

A few setup tasks to make text editing pleasant: You should make sure that if your text editor automatically indents for you, it does so using spaces, not tabs. (An extended rationale for this is given at the beginning of chapter 5; for now, understand that it's important for everyone to be consistent in indenting in certain programming languages.) Ideally, you should configure your text editor so that it inserts four spaces whenever you press the Tab key. In general, you will need to open the settings and find this option. In Atom, for instance, this is done by first pressing Ctrl-, (hold down the Ctrl key and press the comma key) or on a Mac, pressing ⌘-, (hold down ⌘ and press the comma key). Then, select the Editor tab, and change Tab Type to "soft."

Beyond that, one thing that will particularly enhance your programming is having the editor display line numbers. Turn that on in preferences, too, if it isn't enabled already.

[2.3]   Install Anaconda, Our Python 3 Distribution

We will be using Python—specifically, Python 3, the current version of this programming language. Python is an interpreted programming language, so instead of *compiling* a program and producing an executable file, a Python program is generally run in an *interpreter*. (Strictly speaking, Python code is compiled to an intermediate "bytecode" form, and the files ending in .pyc that you will see are signs of this.) The important thing is that as a Python programmer, you don't have to worry about doing anything special to compile it yourself. If you were writing code in C, you would compile your source code to create an executable file, then run the file. To get into technical details once more, you'd actually also need to link the object file produced by compilation into an executable as part of that process, but that's usually done in the same step.

I mention these internal details because computing and programming is full of them—but I'm not trying to intimidate anyone! People are able to be productive as programmers—whether they are doing exploratory or instrumental programming—because of the tremendous amount that is *encapsulated*. When these sorts of things are encapsulated, programmers don't have to constantly worry about them. On the other hand, I thought it was reasonable to mention the difference between *compiling* and *interpreting* code. Sometimes, because Python is interpreted, people will refer to it as a *scripting language*. You can do that if you like, and you could call short Python programs *scripts* if you like. Whatever you want to say about Python, it is a full-fledged programming language.

Although Python is quite powerful and industrial-strength, when you write code in Python, you just need to have your code available as text so that it can be interpreted. There is no need for a separate compilation step; you can either run that code in an interactive interpreter (Jupyter Notebook) or put it in a text file and have the interpreter read that file in and run the program it contains.

Python interpreters are available for many different platforms, which means that many Python programs (and all the ones we'll write here) will work across platforms. The language is widely used both in programming education and in many different real-world situations. Much of what manifests itself to you as Google services, Instagram, YouTube, and Reddit is written in Python—not to mention many stand-alone applications and significant academic research projects. Plus, of course, many instances of smaller-scale exploratory programming, including development of humanities and arts projects.

Users of OS X and GNU/Linux will have a version of Python already installed, and actually they wouldn't need to download and install anything in order to do some work in some version of Python. We won't use the standard interpreter in our explorations, however, because we want to be sure to all use the same version of Python. Also, the interface we're using instead is better for beginners, and possibly for anyone who is trying things out. This system is Jupyter Notebook.

Jupyter Notebook can be installed in the same way whatever your operating system. To install it, you simply download a rather large program that is a distribution of Python and includes the programming language and Jupyter Notebook. This is Anaconda, which is widely used for data science and very appropriate for our purposes, too. Everyone using this book should install Anaconda. This will cause you no trouble, even if you have one or more other versions of Python already installed. You don't need to (and shouldn't!) try to remove any existing version of Python. It's not really a big deal to have different installations of a programming language on your computer. The details that follow about Anaconda installation were accurate as this book was being completed, but specific steps in the installation procedure may change in the future. If something doesn't seem right, check for recent instructions on the Web.

Download and install Anaconda from the following URL, being sure to select the Python 3 version:

anaconda.com/download

It's a large download—and well worth the time and bandwidth.

Choose the installation that is for "Just Me." During installation, you'll be asked if you want to add Anaconda to your PATH environment variable. If you run GNU/ Linux or Mac OS X, check the box. Another checkbox is there to register Anaconda as your default version of Python 3. You do want this, so make sure this this box is checked. This will make it easier for you to work with Anaconda as you use this book and afterward.

After installing Anaconda on your computer—on whatever platform you use— check to make sure you can start your Python 3 Jupyter Notebook. Use the Anaconda Navigator to get started; just run it, as you would any other application. Things may take longer than you expect to start up, but if the installation was done properly, you'll get a new tab in your default Web browser (your browser will open if necessary) that lets you run Python code.

If you'd like an explanation of what's happening here: Jupyter Notebook does something rather elaborate, but very worthwhile. It starts a Web server on your computer and opens a browser to provide access to this local site. You can click on New Notebook to get a Web page that is also a fully functional Python interpreter. You can save and restore these notebooks. While it may seem like a rather involved way to go about

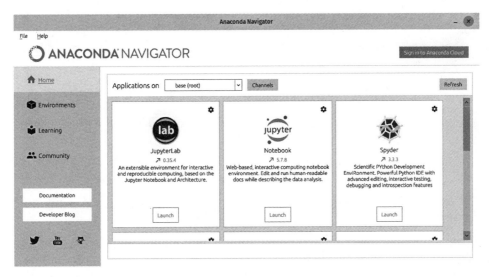

[Figure 2-1]
Anaconda Nagivator will let you launch Jupyter Notebook. Click "Lauch" right in the middle there or wherever you find this option.

providing access to Python, it is a convenient cross-platform method and even allows notebooks to be shared with other people across platforms. It provides a more familiar interface (the browser instead of the terminal window) and lets programmers edit text in a way that is more standard these days.

[2.4]   Find the Command Line

It's important to be able to get around your computing environment. I assume that those coming to this book are familiar with their graphical user interface (GUI) window manager or file manager and can navigate through their folders and find files by mousing around in GNU/Linux, Mac OS X, or Windows. Jupyter Notebook proves a way to program, and to edit programs, that is consistent with the GUI. But for some tasks it is better, or necessary, to use a different, classic computer interface.

Because of this, it's necessary to have a basic ability to use the command line, the text-based interface you have to your same computing environment. You'll need to access your command line and be able to type in commands for the very next section of this book. You will eventually need to know how to move around in the file system, going through various directories. Directories are actually the same as folders, by the way. There is just a different command-line name for them. You'll need to be

[Figure 2-2]
Jupyter Notebook opens as a tab in your default Web browser. In the upper right, click on "New" and select "Python 3" to open a specific notebook in which you can work.

able to view what's in a directory by listing the contents. And it's important to be able to run programs using the command line. Moving through directories, viewing their contents, and running programs from the command line will be all be part of the work undertaken in 7.2, Hello World. For now, be able to start your Terminal program (if running GNU/Linux or Mac OS X) or the Anaconda Prompt (the command line we'll use on Windows). On Windows, you can find the Anaconda Prompt as you would any other application. Clicking Start and typing "Anaconda Prompt" into the search field is one way to do that.

[2.5]   Install a Python Library: TextBlob

We will use one special library, not included with the Anaconda distribution of Python 3. This is a natural-language processing library, TextBlob. You will have to find the command line (as we just discussed) to install this software. Although TextBlob is not used until chapter 13, "Classification," I find it easier to get all the setup out of the way at the beginning, rather than saving any potential frustration for later. If you are doing a short workshop using the book and know you won't be getting this far into the book during your current work, however, you can come back to this when you wish to work with the more advanced chapters.

The easiest way to install TextBlob for use with Jupyter Notebook is using a program called conda; what you type is the same, regardless of platform. The instructions at anaconda.org/conda-forge/textblob explain how to do it. The only difference is that on GNU/Linux or Mac OS X, you begin by opening a Terminal window; on Windows, you open a special sort of terminal named Anaconda Prompt. After doing so, type the following and press Enter:

```
conda update anaconda
```

Agree that you do want to complete the update. Then, type this and press Enter:

```
conda install -c conda-forge textblob
```

Again, agree that you want to complete this installation.

Finally, open Jupyter Notebook and type the following into the first cell to check that your installation was truly successful:

```
from textblob import TextBlob
```

You will probably guess that you need to type this exactly, with uppercase and lowercase letters as shown. At the end of this line, hold down Shift and press Enter (or, on a Mac, Return). If the installation went well, something very anticlimactic should happen: a new cell should appear. This line of code is not supposed to have any visible result. (If you're interested in knowing exactly what it does, it makes the TextBlob object, part of the textblob library, ready for use later on.) If there was a problem with the installation, you will see a message including the text ModuleNotFoundError. First, check to make sure you typed the line exactly as shown. Try it again to make sure the problem is with the installation of TextBlob and not with the way you are testing to see if the installation worked. If the installation wasn't successful, close Jupyter Notebook by clicking on Logout in the upper right and try the installation again, seeking help online if necessary.

There are also generic installation instructions (not specific to Anaconda) located at textblob.readthedocs.io/en/dev/. The recommended installation procedure for Text-Blob (and possibly for other software you are being instructed to install) may also change as new minor versions are released. You may have to check on the Web for the current best installation procedure for your operating system.

[2.6] Install Processing

Downloading and setting up Processing is easy. Just visit processing.org/download, choose the version of Processing 3 for your operating system, download it, and uncompress it. On Mac OS X, you will probably want to place the Processing application in your Applications folder. On Windows, you can put the Processing folder wherever you prefer. You may find it convenient to add a shortcut to Processing on your desktop or to pin it to the taskbar or Start menu. The GNU/Linux version comes with an installation script. Processing is free software and the Processing application is a simple and elegant integrated development environment (IDE) in which you can develop programs. This application isn't very different from a text editor, but it allows you to do several Processing-specific things, including running and stopping programs.

Processing is based on Java and is well-understood as a compiled language. However, you can compile and run programs with a single click. That means the compilation step doesn't weigh very heavily on the Processing programmer.

That said, Processing doesn't have the ability to accept code a single line at a time and evaluate it, the way a Python interpreter does. This facility is used throughout much of the book. That's one of the reasons for doing some work first in Python and then moving along to try out Processing, a capable (and very friendly) system for developing interactive sketches.

[2.7]   Essential Concepts

[Concept 2-1]   Be Prepared

It takes some technical preparation, some setup work, to be ready to start programming. Take the time to get the necessary software installed so that as you continue learning with this book, you can focus on the challenges of programming and computing instead of needing to complete your setup along the way.

[3] Modifying a Program

[3.1]

Even though we aren't trying to learn *everything* about programming or any specific programming language, an important approach of this book, used in almost all of the following chapters, is to build programs up from nothing, to compose them on a blank slate. There is a focus on fundamentals that allow the reader, the new programmer, to build up an understanding of programming that has a foundation.

There is another way to gain experience with computation and programming. You can take an existing program that is somehow interesting to you and start making changes to it to see what happens. And you should try this way as well, for a few reasons:

- You can deal with more complex forms of computation earlier on.
- Some people learn better by referring to an existing program.
- Certain things may be easier to learn working in one "direction" and others in the other.
- Modification provides a way to understand something about working with other people's code.
- This is yet another way of exploring computation, and we should try out any reasonable ones.

While much of the step-by-step instruction in *Exploratory Programming for the Arts and Humanities* is designed to introduce computing fundamentals and help readers develop an understanding of them and build up from them, the free projects in the book provide space for a different sort of exploration ("Let me see what happens when I try this, and if it's interesting . . ."). These projects, a core aspect of the book, invite programmers to try exploring and to follow their own interests, based on concepts that have been introduced ("I've thought of a high-level goal and want to see how to achieve it with programming . . .").

So, let's modify a program. To do this, all you need is a Web browser and a text editor. Your Web browser should be a current-generation browser: I suggest Firefox or Chromium, the free software version of Chrome, but Chrome, Edge, and Safari should work as well. Your text editor needs to be a true text editor. On Mac OS X, you can use Brackets, which is free software, but not the built-in TextEdit, which is more of a lightweight word processor. See chapter 2, "Installation and Setup," for further details on getting a text editor going.

WARNING! Always use a text editor for editing computer programs. Computer programs generally, and Python programs in particular, are plain text files. They are not word processor documents, which are stored in different and more elaborate file formats that preserve typographical and formatting information.

There are specialized environments for programming that combine text editing with other relevant capabilities. Although programs can be edited with a text editor, such IDEs, despite the somewhat imposing nomenclature, can be extremely useful and in some cases can be friendly and easy to use. This book covers how to use the IDE that is included with Processing. This is a well-designed, simple IDE that was created for use in the arts and design. A great deal of our programming, though, will be done within an interpreter. Some programming will be done by editing text files, which can be convenient for sharing final programs (and for turning in assignments). These methods are simple and general, and learning about them helps one explore the possibilities of programming generally, rather than suggesting a dogged focus on systems that are specific to a single language or environment.

[3.2] Appropriating a Page

The first step is to locate a combinatory textual toy in the form of a Web page, several of which are available on my site. There are others you can locate out in the wild, but the ones I will suggest here are packed up nicely with all the HTML, CSS, and JavaScript in a single short file. Here are three simple programs that I wrote and that I can offer from my lexical toy box:

nickm.com/poems/perverbs.html

nickm.com/poems/upstart.html

nickm.com/poems/lede.html

If my quirky approach to text generation doesn't hold your interest, try these historical programs, creative text generators by others from the early days of computing that I've reimplemented for the Web:

nickm.com/memslam/love_letters.html

nickm.com/memslam/stochastic_texts.html

nickm.com/memslam/a_house_of_dust.html

Theo Lutz's "Stochastic Texts" is a particularly good starting point, as it is very simple but (in my view) very well-devised and with some interesting literary engagement. You can find other historical text generators, similarly presented in self-contained Web pages, on the Memory Slam site at nickm.com/memslam. All six of the specific Web pages just listed include computer programs that recombine pieces of text at random, and you should take a look at all of them running online so you can determine which is most interesting to you and (at least to some extent) how it works. The Web versions of these programs are written in JavaScript. They are self-contained, as you can see if you do the following:

1. Go to the page.
2. Select Page Source or View Source so that you see the HTML (which includes the JavaScript). In Firefox, this can be done with Ctrl-U (or on the Mac, ⌘-Option-U), or you can locate the option in the menu, sometimes under Tools > Web Developer. All browsers have a similar option.
3. Select all, usually done with Ctrl-A (or on the Mac, ⌘-A).
4. Copy, usually done with Ctrl-C (or on the Mac, ⌘-C).
5. Open your text editor. Paste all of the HTML into the text editor with Ctrl-V (Mac: ⌘-V).
6. Save that file to the desktop as mine.html—being sure to give it the .html extension. If your operating system is configured to conceal file extensions from you, this is a great time to turn that off so you can see the .html that indicates a Web page and, later on, the .py that indicates a Python file, not to mention the extensions that indicate text files and different types of image files. All the GNU/Linux distributions I know about show the extensions by default. If you use Mac OS X, the option to show file extensions can be reached by clicking on the desktop to activate the Finder and then selecting Finder > Preferences > Advanced. On Windows, open the File Explorer and look at the View tab.

Now, if you open mine.html in a Web browser—you should be able to just double-click it, but if there's some issue with that you can drag the icon into a browser window—you'll see it running. It should be doing the same thing you saw before, but this time will be running right from your desktop, not from the Web. You don't even need a network connection to view this page, having saved it there. Why not prove this to yourself? Close your browser, take your computer off the network by disabling Wi-Fi or unplugging the network cable or by some other (nonviolent) means, and then open the mine.html file again. You can still load the page, because the page is stored locally rather than being on a remote website.

I recommend this exact process, by the way, because it will work across platforms and will provide the initial state of the HTML file. It is possible to download a page in several other ways, but there can be pitfalls in some other cases; this method will work regardless of your operating system.

If you select a page that isn't one of the six I specifically suggested, one that you find out there on the Wild Wild Web, you may end up locating an enjoyable and novel text-combination system to play with. But there are several problems that could arise. You may find that there's no JavaScript code in the page that you downloaded and that the page won't work for our purposes in this chapter. Maybe the system you're looking at is a JavaScript generator, but the code is stashed in another file instead of being included in the main HTML page. Or maybe the system you're looking at uses server-side code in Perl, Python, or some other language, and you don't have access to the source code, only the output. That's why I have suggested you pick one of the six standard examples instead.

[3.3]   Quick and Easy Modifications

The fun can now begin. (Although you might want to put your computer back online so you don't miss an important cat photo that someone posts.) Open mine.html in your text editor once more and find some of the strings that are being recombined: `"It's five o'clock."` or `['Drop', 'box']` or `"sitting on the toilet"`, if you pick one of the first three pieces of mine. Or, if you pick one of the others, look for the data stored in strings (surrounded by quotation marks) such as `'BELOVED'`, `'DARLING'`, `'DEAR'` or `'COUNT'`, `'STRANGER'`, `'LOOK'` or `'SAND'`, `'DUST'`, `'LEAVES'` depending upon which of those pieces you chose to access. Using your text editor, change those words or phrases. Write whatever other text you would like, replacing all (or at least a significant amount) of the text that is there. If you just make a few slight changes, it may be hard to notice the effect, so make your modifications extensive. Save the file and view the new version in your browser (by clicking on the Reload button) to see the result. You probably will want to watch the text being generated for a while to see how the new rules for generation (the rules you developed, by editing text) now work.

Let's explore what happens when we modify a program like this—let's try to do so a bit more systematically. Take a look at line 65 of "Stochastic Texts":

```
subjects = ['COUNT', 'STRANGER', 'LOOK', 'CHURCH', 'CASTLE', 'PICTURE',
```

If we change the word COUNT to something else:

```
subjects = ['HAMBURGLAR', 'STRANGER', 'LOOK', 'CHURCH', 'CASTLE', 'PICTURE',
```

The program still runs. If we look carefully, we can see that we get the word HAM-BURGLAR appearing once in a while and that the word COUNT—which we replaced—doesn't appear at all. That's not much of a surprise, perhaps. But what if we decided to change the very beginning of that line instead? For instance, to this:

```
$#^$&%^#'COUNT', 'STRANGER', 'LOOK', 'CHURCH', 'CASTLE', 'PICTURE',
```

If we try to run this modified program, we'll see that we hosed it. The modified program doesn't run, and no text appears at all. Some more subtle alternations, such as removing one of those quotation marks or one of those commas, can also wreak havoc. But instead of saying "be careful," I'll suggest that you be reckless—just very gradually. Try making one change at a time, viewing the result in a browser, and backing up using Undo if your alteration breaks the program.

A reasonable way of looking at this is from the perspective of code and data. Some parts of this text file are instructions indicating that certain types of computing are to be done. Other parts are, for instance, text strings. If we change one text string to another, there's no real problem. If we overwrite some of the instructions, however, we can mess things up. We almost certainly will, in fact. It's okay; no one is injured in the process. By checking to see which modifications are changing data and which are changing code, you can start to understand what code and data are. We could go on to make more fine-grained distinctions, but distinguishing between code and data is a very helpful start.

Feel free to keep messing around with the data you've identified, those text strings. What if you put the same word in several slots? What if you systematically replace a whole set of words, so that instead of 'COUNT', 'STRANGER', 'LOOK', 'CHURCH', 'CASTLE' (which Theo Lutz took from Franz Kafka's *The Castle*) the terms used are 'WHALE', 'SEA', 'MAST', 'HARPOON', 'LEG' and other words prototypical of Herman Melville's *Moby-Dick*? The results of a modification do not always have to be silly or funny; they can provide a different type of seemingly endless and disturbing recombination, as with this shift from more institutional repetitions to nautical ones.

After trying that out, you can see if some ways of changing the code turn out to have interesting results, rather than just breaking the program. What happens if you change the number of strings, removing or adding some of them from an array? Can you make text appear more rapidly or more slowly? Let's consider how that would be done on any of the six standard pages.

You can see that new text is being produced about every second or every few seconds (in other words, not as quickly as every millisecond, and not as slowly as every minute). The rate is probably going to be represented as a number; that is how all of us tend to measure time. So, find where there is a number in the JavaScript code (not anywhere else in the HTML, but just in that section beginning with <script>) and make a dramatic change to that number. Run the program again. Did the rate at which text is being produced change dramatically? If not, undo your change and find another number to change. There aren't very many in the code.

By the way, if you see a number such as 2 in a program, it is quite likely to be a sort of data as well. Strings are not the only type of data. But this leads to an interesting thought: What would happen if you changed 2 to 0 and the program, which had been dividing some result by 2, was now trying to divide by 0? That's right—you would hose the program again, this time just by changing data. It's possible for data changes to disrupt the working of a program, but programs are generally more robust to arbitrary changes in data than to arbitrary changes in code, as you can see when you modify the program you have chosen.

Can you simplify the program (perhaps having it do something different and simpler) by removing parts of it? Can you build up from the simplified version, with reference to the original, and create your own structures of combination?

Making modifications of this sort doesn't require understanding how JavaScript works in any detail or depth. You can change strings around and see the result without worrying about how the program is accomplishing that. And even removing or adding strings doesn't take much background. What if you make a change that breaks something totally, such as adding a jumbled bunch of characters to the JavaScript? It's not a big deal. Just use Undo in your text editor and try again. Or, if you can't undo enough to get things working, just start over with the original page.

[Free Project 3-1]   Modify a Simple Text Machine
Do this project three (3) times, using two (2) different Web pages as your starting point. Start with a particular combinatory textual toy in HTML with JavaScript—a simple one, although your selection doesn't have to be one of the six listed previously. Modify the program so that its output is somehow substantially different. It could

contradict the original system, for instance, or the tone could be completely different, or the system could produce unrelated output. Beyond replacing some text with other text, see if you can make changes to the way the system functions. For instance, if your system generates sentences, can you change the syntax of the sentences it outputs—adding adjectives or moving parts of speech around—rather than just changing the lexicon? For this exercise, restrict your work to the JavaScript code (in the <script> element) rather than modifying the HTML. Modifying HTML can be fun and rewarding too, but our focus is on programming, not on structured documents or their appearance.

[3.4]   Share and Discuss Your Projects

As I've tried to emphasize, it's important to do the free projects and to share them—beginning with this one. In a classroom of any size, each programmer should share his or her modified text generator. The easiest way to do this is by having programmers read aloud from their system's output. Every time that I've asked people to do this, I've heard an interesting diversity of outputs that vary in attitude, subject matter, and register, among other aspects. Sharing work in this way also shows that everyone has the ability to create new programs, in this case by modifying existing code.

Programmers who aren't in a formal class should find opportunities to share their work with friends, by showing the system to others in person and by putting the page online somewhere and having people look at it and discuss it online. Of course, it's better to share your project with people, online and off, who have some sort of experience with computational art and literary work, however informal. If your friends are expecting a video, a meme, or some static text, you may not get useful responses. But you don't need to find expert critics to get useful responses. As long as people are open to the sort of work you're doing, they can have good ideas and comments.

Ideally, programmers learning with this book should get feedback on at least one of their attempts at each free project. In a few cases, this may come mainly from an instructor or teaching assistant because not all classes will accommodate workshop discussion of all projects throughout the semester. Whenever possible, however, it's ideal to share and discuss work as class. For those not in a class, even seeking a few informal responses can be valuable, as others may point out interesting aspects of a project that the programmer overlooked.

To be able to look over your own work again as you continue through this book, I suggest you move the mine.html file off the desktop and into a folder that you have dedicated to your *Exploratory Programming* work. You can create folders within that folder for each chapter if you like, or use a file naming system, based on the way exercises and free projects are numbered. To share your work with others and facilitate discussion about it, you have a variety of options. You can simply attach files to email and send them along, although I recommend compressing a file into a ZIP file because some email programs will process and mangle HTML attachments and other text files. You can become part of a community such as codepen.io and use the online editor there to put a page up for others to view and discuss. You can also place an HTML file online more permanently, having it hosted so that anyone with access to the Web can see it. Some free online services exist for this, but because the ones that are up and running tend to change every few years, I'll leave you to search for the ones that suit you best.

[3.5] Essential Concepts

[Concept 3-1] Programming Is Editing Text

All you need to do to modify a program—to create a new program based on a previous one—is edit a text file. Materially, that's like using a word processor, except even simpler. Compared to editing video, practicing a martial art, or undertaking tablet weaving, this is easy. It can done in many different places, using a standard notebook computer. Programming is just writing a special kind of text.

[Concept 3-2] Code and Data Differ

Can you distinguish between instructions (*code*) and particular values (*data*) in the program you modified and in another JavaScript program you find and consider? How could you test your theory that some part of a JavaScript program is data?

[4] Calculating and Using Jupyter Notebook

[4.1]

Now that we've developed a program by starting with a similar one, we'll see what we can do to develop some snippets of code from scratch. We'll start with some very, very tiny snippets of code that should look quite familiar.

[4.2] Calculations

Start Jupyter Notebook and click on New in the upper-right corner. Then, under Notebook, select the Python 3 option. This will open an interpreter, or notebook, for the Python programming language, right within your Web browser. After you have done this, just type the following simple expression into the text field (also called a *cell*) at the top, the one labeled In [ ]:

```
2 + 2↵
```

The ↵ symbol at the end means that you are to press Shift-Enter—that is, hold down Shift while pressing Enter (or Return on a Mac). This indicates that you're done with the whole block of code being typed into the Jupyter Notebook cell. Once you've typed this, you will get the answer that you expect, 4, and a new cell will appear in which you can type something else. You have just used Python, and Jupyter Notebook, in a simple way: as a calculator.

Next, type this in the current cell, with Shift-Enter at the end:

```
17.35 * 1.15↵
```

The result here is the amount you might pay if you have a $17.35 bill, and if you think you should tip on the tax that is included in that, and if you wish to tip at a rate of 15 percent—quite stingy in my part of the United States. You can see from this

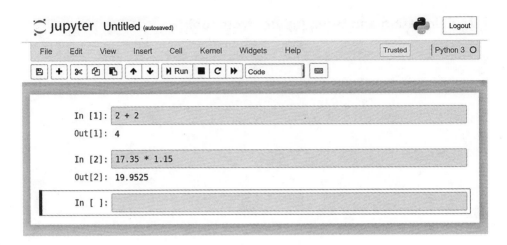

[Figure 4-1]
The first two code snippets in the first two cells of a particular notebook, opened in Jupyter Notebook.

computation that in this situation, leaving a twenty would be suitable. I'm not usually going to give away what the answers are, but to orient you to Jupyter Notebook, Figure 4-1 shows what it will look like after you've typed in these two expressions in the first two cells.

The cell labeled In [1]: holds the first input. What results from evaluating this expression appears next and is labeled Out[1]:. The following cell with In [2]: before it is cell 2, with the input first and the output following it. The cell labeled In [ ]: is the current cell, which is ready for new input.

The point of these simple computations, getting the computer to do arithmetic for you, is to show that the types of computing ordinary people do every day can also be done by a computer. But let's be even clearer about this: these sorts of arithmetic expressions are actually valid bits of Python code. It's not very useful to think of them as computer programs by themselves, but they can be included in various statements in a program—in conditions and assignments, for instance. So if you know how to type "2 + 2", then you *already* know how to type some valid Python code.

In the current cell, the one labeled In [ ]:, try an even simpler expression:

2↵

You'll see that Python knows the answer to this one, too: 2.

The computer is essentially a symbol-manipulating machine. While numbers don't have to be the symbols, the computer has historically been understood, in culture, or at least in English-speaking cultures, as a calculator. Some early computers were even called calculators rather than computers: the Electronic Delay Storage Automatic Calculator (EDSAC) and IBM's Selective Sequence Electronic Calculator (SSEC), for instance. This primary identification with computing and calculation is not the case in every culture or language: in French, the standard word for computer, *ordinateur*, literally indicates something that orders and organizes. However, *calculator* and *computer* have been recognized as equivalent terms in French as well, and the calculating function is recognized as a major one in computer-using cultures.

Given this, it's no surprise that modern programming languages make it easy to calculate. What is happening when you type 2 + 2 and get 4 or even when you type 2 and get 2? In these cases, you typed an expression, and the interpreter replied with what the expression evaluated to; it did the arithmetic and provided the number that is the answer. You should once again type the following:

```
2+2↵
```

You'll see, again, that the expected answer is provided. By the way, you can type this in exactly as is (without spaces), or you can add in a space on either side of + (as at the beginning of this chapter), or you can add as many spaces as you desire. From here on out, I'll always place single spaces between parts of arithmetic equations for stylistic reasons, not because it's necessary in terms of syntax. Style exists in programming just as it does in writing, not only in terms of how computations are done but also in terms of how they are written (or typed). This sort of style is related to programming conventions. The dual nature of programs is important here: Programs are both run by computers (which couldn't care less about such stylistic niceties) and also read, modified, and written by people (who can benefit greatly from good style).

Not everyone agrees about what good programming style is, and some stylistic slips are widely considered worse than others. You can choose to type a+b or a + b; few people are likely to complain about the difference. In every case, I've tried to follow good programming style conventions in this book. You should know, though, that just as there is a house style for particular publications, different companies use and often enforce different style guidelines for their programmers. This can be the case with other sorts of collaborative projects, too. Also worth noting is that there are standard style guides, not just house styles, for many programming languages. In the case of Python, the standard style guide is *PEP 8*, discussed further at the beginning of the next chapter. This guide explains how to format code for human readability, much in the same way that the *MLA Handbook* explains a standard format and style for scholarly writing in the humanities.

Of course, you can subtract as well in Python—and please do:

```
5 - 25↵
```

Division is of course possible in Python, too, and indicated using the / or slash. For that matter, there are other operators that, for instance, work on Boolean values (True and False) and work on bytes, treating them as a sequence of eight on or off bits. For the moment, however, it's adequate that we use addition, subtraction, multiplication, and division. That is also enough to let us create more complex arithmetic expressions that use more than two numbers and more than a single operator as well.

Putting aside the discussion of many other operators and of the expressions that result from using them may seem unfair. Or the existence of all this complexity may cause anxiety. But my choice to bracket certain features of a programming language (and to make a quick mention of them at this point) is not capricious. It reflects a very basic principle in computing. It is one aspect of *abstraction*, which can be understood, in part, as carefully controlled ignorance. Abstraction involves selecting whatever is of interest and leaving aside that which is unnecessary at the present time. That's why in a scientific paper, an *abstract* can be one paragraph long and can nevertheless manage to make the main point, while the text that follows it is usually many pages in length.

When humanists enter a library in search of a particular book, they do not typically fall to their knees, begin to weep, and shield their eyes from all the books in the stacks. (I haven't ever seen this happen, anyway, and I also haven't done this myself, even if America's libraries are full of tears, according to Allen Ginsberg.) "How can I ever go and select the *one* book I am interested in," they don't say, "without reading *all* the books that are here?" No—instead, they simply go past a large number of those books and locate the particular book that is of interest at that particular point. Then, if these researchers are anything like me, they read only a part of that book. A researcher might have taken a look at nearby books on the shelf (as books in libraries have been meaningfully arranged, or *colocated*), but the inability to read everything is no obstacle to research.

There are some things in programming, as in other areas, that it's worth knowing about thoroughly. More complex arithmetic expressions are of course among them, which is why these will be discussed later. But it's also important to grasp the necessary and fundamental aspects of programming without being overly concerned about learning *everything* about a language. An attempt to learn to program by exhaustively advancing through the workings of every arithmetic operator, and then every Boolean operator, and then every bitwise operator, and then every Python keyword, and then the interface of every built-in object, and so on, and so on, would be rote, tedious, and detached from any real purpose.

Any user who performs the simple inputs presented earlier already demonstrates an ability to do work in Python, to employ the Python interpreter for a useful purpose. Sure, it's not a highly impressive use of the language, but it nevertheless shows that the work that can be done in a calculator application, app, or widget can be carried out in Python. In fact, it can be carried out just as easily. Indeed, you can use Python to compute grades or taxes instead of using a calculator application or an old-fashioned hardware calculator. I often open a Python interpreter to use it for a quick calculation.

As for trying to learn every aspect of Python (or of programming more generally), the approach we will take here is to learn enough to understand the fundamentals and enough to usefully explore topics that matter to us. Those who develop interest in specific media or specific aspects of Python will want to develop more ability, and expertise, as well; what can be learned with the help of this book should provide a good basis for such further work.

[4.3] Encountering an Error

We've seen that 2 + 2 evaluates to 4 and that 2 evaluates to 2. Now, let's see what happens when you type in the following:

```
2 +↵
```

A message is emitted, ending with `SyntaxError: invalid syntax`. Here we witness the dreaded syntax error. According to some, this error is—gasp!—the bane of the programmer's existence, the most profound fountain of frustration.

Is it, truly? I would suggest instead that the syntax error message is the friend of all programmers, whether they are novices or deeply experienced. The Python interpreter in this case simply explains that we do not have a *valid* line of code or arithmetic expression: 2 is a valid expression; 2 + 2 is a valid expression, but 2 + doesn't mean anything in standard arithmetic. You might associate with it something along the lines of "two or more" or "two more," but it does not follow the syntax of arithmetic expressions. Just as it is not possible to withdraw 2 + dollars from one's bank account (whether you go to an ATM or talk to a human bank teller), a computer cannot evaluate 2 + as if it were an arithmetic expression.

The error message here is actually offering *assistance* to the programmer. It is showing where what seems to be valid Python code ends. Consider this slightly longer text that is also not a valid arithmetic expression, along with the message that results, which will be something like this:

```
(2 + ) * (3 - 1)↵

  File "<ipython-input-7-02f34210bdd2>," line 1

    (2 + ) * (3 - 1)

         ^

SyntaxError: invalid syntax
```

The last line says that there's a syntax error, that what we typed is *invalid*. The line above has just a single character, a caret symbol, in it. It's being used as an arrow to point to the first) in the line above. And the line above that, the first line of output, indicates that what we're looking at was typed into Jupyter Notebook. Specifically, it's from `<ipython-input-7-???>`. Jupyter Notebook used to be called iPython Notebook, and that's essentially why `ipython-input` appears here. The 7 indicates that this is the seventh cell or input, which is where you will have been if you input exactly the code I've provided. (It's fine if you tried out some additional expressions—seeing how division worked, for instance.) The `???` is some gobbledygook that really isn't of relevance to us; let's use our *abstraction* skills to forget about it. After that there's an indication of `line 1`. It's sort of obvious that the error is on line 1, since we've only input a single line here, but this line number can be quite useful when there is more than one line.

Here we have something that is *almost* a valid expression. And the error message points precisely to the problem. Now, it doesn't say "put a number here!" and in fact a number isn't the only valid thing to include at that point. But we at least are shown where the problem is. If we did insert a number right before the character the caret is pointing to—for instance, the number 2—we would actually be adding two numbers together at that point, and the overall expression would be valid:

```
(2 + 2) * (3 - 1)↵
```

We had some text before; with a very slight change, a one-character change, we have an arithmetic expression. The syntax error message pointed directly to the problem. While problems are not always so clearly indicated by error messages, in most cases they do indicate the first place in the code where something is amiss. That may not be every place where an error is present, but it is the first place, and the best place to start.

Consider an attempt to add five and five together and to multiply this by eight minus one. The way to express this in mathematical notation is $(5 + 5) \times (8 - 1)$. We use an asterisk (*) for our multiplication operation when we're using a computer. Let's now see what happens when this is typed in with two syntax errors, the omission of the second 5 and the 1:

```
(5 + ) * (3 - )↵

  File "<ipython-input-9-787a9fd4bc10>," line 1

    (5 + ) * (3 - )

          ^
```

SyntaxError: invalid syntax

The error message indicates the first place where the expression is not syntactical. It's where a 5 has been omitted. When you add a 5 (you need to be typing this in!), there will be another error message indicating the second problem point:

```
  File "<ipython-input-10-ec89113804fc>," line 1

    (5 + 5) * (3 - )

                  ^
```

SyntaxError: invalid syntax

This shouldn't be all that frustrating. A term has been omitted and there is no way for the computer to know that the number 1 was intended. The best that the computer can do (specifically, the best that can be done by our Python interpreter, which is Jupyter Notebook) is to point to that gap, that lacuna, to indicate where what looks like the syntactically valid code ends. Having noticed that, it's easy for the programmer to include the missing digit and repair the expression.

It is common for in-development computer programs to have more than one error. The appropriate way for a programmer to handle this is to determine where the first error is and fix it. Then, once again, determine where the first error is now located, and fix it, and so on.

Couldn't in-process programs have quite a few errors? They could—and there are two ways of dealing with this.

First, a programmer can avoid writing long programs or long stretches of code without trying to run (or compile) what has been written so far. Write as little as possible, modularly, and continually check to see that it's syntactically correct. Expert programmers who seldom slip up syntactically nevertheless adopt this practice, asking the computer to help them ensure the validity of their code.

Second, a programmer can understand that discovering syntax errors and repairing them is just part of the development process, as is writing code to begin with.

Python's SyntaxError, and the way the error message indicates where the problem lies, isn't some strange feature of this programming language in particular. Any programming language will indicate when code can't be interpreted or can't compile. While some languages are better at pointing to the specific part of the line and others are more cryptic, all of them use error messages to help programmers locate and deal with invalid code.

[4.4]   Syntax and Semantics

In the previous example, I explained how error messages identified two syntax errors, one after the other, and how they helped the person developing the expression to repair them.

However, there is a problem with the valid expression that was developed as a result. According to my original statement, "Consider an attempt to add five and five together and to multiply this by eight minus one," the goal was to compute $(5 + 5) \times (8 - 1)$. What was developed through this process of correcting syntax errors actually computed $(5 + 5) \times (3 - 1)$—and *that* expression has a 3 where there should be an 8. The answer, then, is computed to be 20 while it should be 70.

This is an error, obviously, but it isn't a syntax error. The Python interpreter correctly accepts that (5 + 5) * (3 - 1) actually is an arithmetic expression, because according to arithmetic, there's nothing wrong with subtracting 1 from 3 instead of subtracting it from 8. This is formally just fine; it's an arithmetic expression either way. The problem is not that the expression is invalid; the problem is that it isn't the *right* expression. The error is in the *semantics*. It is the meaning that is wrong.

Another example: standardized numbers have a particular form. In the United States, basic postal codes (zip codes) are five digits, while phone numbers have a three-digit area code followed by a three-digit exchange followed by four more digits. If we assume that we are within a particular area code, as people used to do, it would be typical to express a phone number in seven digits. In the movie *Taxi Driver*, a Secret Service agent asks Travis Bickle, played by Robert DeNiro, for his zip code. Bickle immediately says, "610452." When the Secret Service agent tells him that this is six digits, Bickle says that he must have been thinking of his phone number. He's switched from one syntax error to another, from giving one formally wrong answer to another formally wrong answer. In this case, the Secret Service agent must be pretty sure that Bickle is just making up a number, given those syntax errors.

Email addresses also have a syntax, of course, and someone who claimed, either to a person or a Web page, that their email address had a # in it but no @ would be easily exposed as providing an invalid email. To determine if an email address or phone number actually functions (whether the phone line is active, whether an email server is at that domain and has an account of that name) involves further steps. And, beyond that, additional work is needed to determine if the email address or phone number actually belongs to the person who has provided them. But the point with regard to syntax, and to formal validity, is that it's possible to initially check to see whether the *form* of something like an email address is valid; if it isn't, there's absolutely no way a phone number or an email address or whatever else will work.

When developing a computer program, the goal is always twofold: on the one hand, develop a working program; on the other, have the program work in the intended way. If the text that has been typed isn't valid—if there is even one syntactical problem—then it isn't going to work. Strictly speaking, one could say that it isn't even a program. It's rather confusing, though, to speak of a file that becomes a program and ceases to be a program as minor changes are made to it. So, I will refer to *valid programs* when I discuss those that will run and use terms such as *programs-in-progress* if I wish to specifically include texts that are being edited and that may not be valid programs. The important point here is that even when a valid program has been developed, this working program may or may not operate as intended.

Programming in an exploratory mode doesn't exempt one from having programs that work as intended (or not). The nature of this proper function certainly may be different than in a banking or military system that is developed by specification-based programming. The programmer developing a creative project may be willing to follow what was apparently a wrong turn, for instance, endorsing what was originally a mistake. But if the programmer intended a visual effect and instead developed a valid program that drew nothing, simply leaving the screen blank, this would almost certainly be an unhelpful wrong turn. If a humanist developing an analytical program is not successfully reading in the data that is supposed to be considered, that too constitutes an essential problem, resulting in a program that isn't going to work as intended.

The aim of this book is to empower artists, humanists, and others with the ability to program and explore. For anyone to do so, it's important to be able to develop programs that are both valid and intentional:

Valid programs can be interpreted and/or compiled. They are formally correct.

Intentional programs are always valid, but they have an important additional property: they do what the programmer or programmers want them to do.

The intentional program is not meant to be a fine philosophical concept. It just distinguishes a program that needs to be further revised from one that accomplishes the current goal. After developing an intentional program, a programmer may certainly choose to set new goals and continue development; this would indeed be typical in exploratory programming.

[4.5] A Curious Counterexample of the Valid and Intentional

As presented, formal validity and working according to intention should seem reasonably clear-cut. The way they have been portrayed so far is generally accurate with regard to programming, and certainly for programming in Python. When these concepts get into the wild, worldwide context of other forms of coding, however, what results is not as easy to classify.

HTML (Hypertext Markup Language), a way of structuring linked documents, is an excellent case in which formal validity and intentional functioning are both meaningful concepts but work differently. Whereas a JavaScript program is a program, a page of HTML, by itself, is not. HTML does not encode instructions for computation, which is why it is called a *markup* language rather than a *programming* language. Like a phone number or email address, however, HTML can be valid or invalid.

A Web browser is supposed to be liberal and allow much more to be rendered than is actually syntactically correct. For instance, by definition a link cannot run across two different paragraphs in HTML. The valid way to mark up a page that is supposed to present a link of this sort is to create one link running to the end of the first paragraph and another link (to the same URL) beginning at the start of the next paragraph. That said, browsers are designed to be forgiving, or liberal, and will render an invalid page whenever it's possible to formulate some output, even when this is based on a loose interpretation of HTML.

On the other hand, one is supposed to be conservative in writing and posting HTML. Although browsers are forgiving, the creators of Web pages are *supposed* to write clean, valid HTML. That's a nice idea. But millions of people write HTML or cause it to be generated; only a few write widely used Web browsers. Those millions of people want to get their Web pages up quickly. As a result of these and other factors, one is hard pressed to find examples of valid HTML on the Web.

The rule that motivates this "liberal" and "conservative" behavior predates the Web, but certainly applies to it. Jonathan Bruce Postel, in one of the canonical "Request for Comments" documents specifying the behavior of the Internet (RFC 760), established what has come to be called Postel's law. In this 1980 document, he wrote that "an implementation should be conservative in its sending behavior, and liberal in its receiving behavior." Later, in RFC 1122, this was rephrased as "be liberal in what you accept, and conservative in what you send."

When a Web developer is interested in HTML that will be readable years or even decades from now, on whatever browsers might exist in the future, it is sensible to consider the validity of the markup used. The World Wide Web Consortium offers a validator for HTML at validator.w3c.org, and there are many good stand-alone validation tools as well. Perhaps it seems that there is little practical need for such validity checks, particularly when working under deadline to pop pages out of the hopper. After all, it's easy to view a page in a browser and see whether it looks right. The problem with this reasoning is that a browser is forgiving. Other existing browsers may not be forgiving in the same ways, and future browsers on whatever is to follow mobile phones and tablets may not be forgiving the same ways, either. Because of this, a validator is great for learning HTML in the first place and for developing hand-coded Web pages that are robust and archival. If we were covering markup in addition to programming in this book, I would send you to the validator to get the same sort of helpful responses that Python's syntax errors offer.

Given the unusual way in which Web browsers and HTML work, it's possible to have an *intentional* Web page, one that opens up in a particular browser and is structured in the way it is supposed to be, which is at the same time not a *valid* Web page. Of course, if an author follows Postel's law, they will validate all Web pages created, being conservative about what is sent out on the Web server. But on the Web, validity is just an option. When you are programming, validity is a requirement.

If you haven't worked with HTML, this should offer a glimpse of how the Web works and its (somewhat odd) relationship to validity and intentionality. If you have written and edited HTML, I hope this explains one important way in which writing HTML and programming in Python and other languages is different. Although it's possible to develop an invalid document that is "good enough" for current browsers, an invalid Python or Processing program won't work. Formal correctness is essential. Fortunately, it's not impossible or even difficult to attain valid code, with the right mindset and the help of the interpreter and its error messages.

[4.6] Using Jupyter Notebook

You now have a general sense of how Jupyter Notebook works because you have typed some arithmetic expressions into it, seen them evaluated, and seen errors produced when invalid expressions were typed in.

Jupyter Notebook is, like a notebook, a good space for sketching, making notes, and trying out ideas. Unless used properly, however, it can obscure what code you have run and (as you will soon see) what values have been assigned to what variables. That's because you can overwrite the contents of a Jupyter Notebook cell with something entirely different from what you placed in that cell before. This can be deeply confusing.

There is a simple way to avoid this problem, but it requires attention and discipline on the programmer's part.

As you use Jupyter Notebook, don't ever replace the contents of a cell once it has successfully run. If there's an error in certain code, it's fine to edit the code in that cell and fix the error. Once a cell has run, however, you need to move down to the next cell and start working in that one. Otherwise you will lose the trail of computation, which can be perplexing for anyone, but particularly for those new to programming.

I suggested at the end of the last chapter that you make a folder for your work with this book, which should have your mine.html file in it. You can also give your Juypter Notebook a specific name using File > Rename (if you don't choose one, it will be called Untitled.ipynb) and move that notebook to wherever you like. I suggest keeping these files organized by date and/or chapter within your *Exploratory Programming* folder so that you can easily look back at what you've done.

[4.7]   Essential Concepts

[Concept 4-1]   Arithmetic Expressions Are Snippets of Python
Understand that expressions such as 2 + 2 can be evaluated by a Python interpreter because they are valid bits of code that can be usefully incorporated into programs. Be able to write ordinary arithmetic expressions, using parentheses to be clear and when necessary. Simply put, be able to use Jupyter Notebook as a calculator.

[Concept 4-2]   Syntax Errors Help Us Attain Formal Validity
Be able to explain what a syntax error is and some basic ways in which an error message can help a programmer fix nonworking code. How can an error indicate which specific part of a program-in-progress has a problem? What will happen if there is more than one problem?

[Concept 4-3]   Valid and Intentional Are Different
Explain the difference between a phone number, arithmetic expression, or other piece of code that is *valid* and one that is, as described in this chapter, *intentional*. Can you

have an intentional Python program that isn't valid? Can an email address be valid but not intentional? Can a Web page be intentional but not valid?

[Concept 4-4]   Always Proceed Downward in Jupyter Notebook

Instead of replacing successfully run code with other code and overwriting the contents of Jupyter Notebook cells, keep moving to the next cell as you compute. This will leave a clear record of what computation has been done so far in your session.

[5]   Double, Double

[5.1]

A program that prints "Hello World" is a traditional first computer program. In chapter 7, "Standard Starting Points," I discuss this category of introductory programs, "Hello World" itself, and a few other starter programs of this sort, explaining what they have to show us about computing and culture. For now, I offer a code snippet called "Double, Double" as our first program to be typed in and used. (To be precise, it's not a complete program, but is our first function, one that we can use within Jupyter Notebook to produce a result.) Artists and humanists may appreciate that "Double, Double" is a reference to the weird sisters in *Macbeth*, although I can't claim that I'm particularly special because I made a cultural reference in my program. Even the venerable "Hello World" could be seen as a reference to Miranda's exclamation "O brave new world" in *The Tempest*. If you don't buy that one, consider that the source code from the Apollo Guidance Computer's software has routines that programmer Don Eyles referred to using the names of characters from *Hamlet*, Rosencrantz and Guildenstern (Dobson and Mosteirin 2019). Many computer programs found "in the wild" make different historical, literary, and film references, among other sorts of references. Computer programs are cultural artifacts, so why shouldn't they make such references?

The meanings of even a simple phrase can be complex, too. Perhaps the name of this program will particularly resound for Canadian readers, since "Double, Double" also is a request for a coffee with two sugars and two creams at Tim Hortons. Or it may please readers in the vicinity of an In-N-Out Burger, where among the nonsecret menu options they can find the Double-Double.

[5.2]   Type In the Function

The following Python code defines a function double(). You need nothing except a
Python interpreter to type it in and see it work. So, just open Jupyter Notebook and
type this directly into the first cell:

```
def double(sequence):

    result = []

    for element in sequence:

        result = result + [element * 2]

    return result↵
```

After you type the first line in Jupyter Notebook and press Enter, you'll see that the next
line is magically indented for you. (It's not really magic, but rather the result of typing
a colon at the end of that first line.) Keep that indentation as it is; don't add or remove
any space. After you type the third line, you'll see that there is more indentation provided. Keep that indentation just as it is and continue by typing r and then the rest
of that line. When you press Enter at the end of that line, you'll see that there is more
indentation than you need. Backspace once to move back a level. (In Jupyter Notebook,
the automatic indentation is four spaces, pressing Tab gets you four spaces of indentation, and removing indentation with Backspace takes away four spaces.) Then, type
the last line and press Shift-Enter to complete this snippet of code. You'll just get a new
cell—no value results, as it does when you type in an arithmetic expression. But you
will have defined a function.

   At first, we'll focus on exploring what double() does and how to use it, rather than
detailing how this code actually works. Don't worry; we'll get to all of those details
later, in chapter 6, "Programming Fundamentals." For now, though, it's worth mentioning that function definitions like this one are made using a particular sort of template. The way to define functions is formulaic, at least in its overall framework. A
function definition will always look like this:

```
def ____(____):

    ____
```

The first blank contains the function name, which can be almost anything—it shouldn't
be a Python keyword such as def (those words are reserved for special uses), but can be
anything ranging from double to mean to illuminate to asdfpurq. Of course, some of
these names are likely to be more clear than others, but understand for now that the

choice of a function name is up to the programmer. It isn't a particular required word, as def is.

The second blank contains zero or more arguments. Functions don't have to have an argument at all, and may have more than one. We'll see what those cases look like soon enough. In the current case, double() takes one argument, which is labeled sequence; that's the name of this argument.

Because the parentheses are conceptually attached to the function name, it's proper to write double() instead of double (), and it's proper to always attach that left parenthesis to the function name when defining the function and when calling it. In this book, function names are always shown this way, and in any well-written code they will be. Remember that you're not making some sort of optional, parenthetical comment after typing double; you should place the left parenthesis directly after the word. Even if your code happens to work with a space before the left parenthesis, it isn't stylistically correct to have space there and it could be confusing.

The third blank, which might represent more than one line of code, is the function body—whatever code is executed when this function is called. In the present example, the function body is four lines long.

After their names, functions always have parentheses when they are defined and when they are called, and if they require one or more arguments something will be in those parentheses. At the end of the first line of a function definition, there must be a colon to meet the syntactical requirements, to abide by the formulaic way in which functions are declared. But the rest is up to the programmer.

Indentation is not decorative in Python. It is meaningful—part of the *syntax*. Indentation indicates special blocks of code, including function definitions and code that runs within loops. After the first line of double(), beginning with def, the code is indented, and after the next line, beginning with for, it is indented again. If you were typing this function in a text editor, you would need to add that indentation yourself. The way to do it is with four spaces—not two, three, or any other number of spaces, and certainly not (gasp!) a tab. If you copy the code snippet you just entered from Jupyter Notebook and paste it into a text editor, you can see that the indentation you were automatically provided consists of four spaces (for the second line) and eight spaces (for the last).

As you work through the exercises in this book, don't ever add a tab character into Python code (or any other code). If you are using your text editor and have it set up so that pressing Tab adds four spaces, fine, press the Tab key to indent; just don't add the tab *character* itself. Now, in Python you are officially (by the language itself) allowed to use either type of indentation, tabs or spaces, and the *consistent* use of either will result in programs that are valid and work. Mixing tabs and spaces, however, causes particular headaches for programmers as it creates errors that are literally invisible. When trying to collaborate and combine code written by different people, the problem is worsened. The issue of which type of indentation should be used can be a heated one, even though *PEP 8—Style Guide for Python Code*, coauthored by the creator of Python in 2001, holds that "for new projects, spaces-only are strongly recommended over tabs." Ultimately, this matter is similar to the question of whether to drive on the right side or the left side of the road. Either side is fine, but what really matters is that everyone agrees on one and sticks to it. In the country of *Exploratory Programming for the Arts and Humanities*, we will drive on the "spaces" side of the road, which is also where Jupyter Notebook drives and where Guido van Rossum, the creator of Python, suggests we stay.

The spaces versus tabs issue illustrates that there are some standards and norms determined by convention, not by changing the workings of a programming language to enforce that only one way is possible. Either way is syntactically correct, but programmers do well to use a consistent style and stick to a particular way. If they don't, they risk confusing and inconveniencing themselves. Writing `double ()` instead of `double()` is a stylistic mistake of a certain sort, but mixing spaces and tabs is a stylistic mistake that is likely to have consequences for getting a program working and maintaining it. The consequences are clearest in Python, where indentation is essential to the workings of programs. But there's also no reason to use tabs in any other programs in languages discussed in this book—Processing or JavaScript.

[5.3] Try Out the Function

After you type in "Double, Double," you will be able to use this function and see what it does. Specifically, you will now call or invoke the function by giving it a sequence as an argument. One type of sequence in Python is a list, which can look like this: `[1, 10, 5]`. That particular list is one with three elements, the first one 1, the second one 10, and the last one 5. The numbers are arbitrary; I'm just providing these so that you and everyone else using this book will try out `double()` in the same way at first. I've started with small, positive integers (numbers without decimal points) that you will be able to double in your head so that you can check to see what the function is doing. To invoke the `double()` function on that list, just type:

```
double([1, 10, 5])⏎
```

If everything was typed in as shown here, the answer that comes back—the values that the function returns, which are displayed in the interpreter—should be shown in a similar format. To deter those who are trying to read this book without using a computer and typing things in, I'm not including the output. I will usually omit the output. You will need to type in the `double()` function and then type in the line provided to see what results. After you do, try this and a few other lists of numbers:

```
double([-42, 0, 42])↵
```

Negative numbers and zero should be doubled just like positive ones. Try a longer list with four, five, or ten numbers in it. You can also check to see what happens if a sequence has only one element. Will the code still work?

```
double([24])↵
```

No problem. What if the list has *no* elements? Try typing the following line in—but this time, *wait* before you press Shift-Enter:

```
double([])
```

Take note: I omitted the Shift-Enter symbol—↵—from the preceding line to indicate that you should not press that key combination yet. I'm asking you to imagine what will happen before you see what happens. What do you expect? Will the program crash? If so, what sort of error will it generate? If not, what will be displayed? Formulate a hypothesis of some sort, commit to it, and after you have done so, press Shift-Enter:

↵

Interesting . . . perhaps? There are the brackets that are typical of a list, but with nothing between them, as with the argument that was passed in. Like the sound of zero trees falling in the forest, this result is the empty list—the list of zero elements.

Now, what if we don't give the `double()` function a list at all? What if we remove the list entirely from between the parentheses? Again, *without* pressing Shift-Enter at the end, type:

```
double()
```

What do you expect to happen this time? Will the program crash? If so, what sort of error will it generate? If not, what value will it return? Whatever your guess, be sure to have a guess of some sort.

↵

People can only operate computers because they have some mental model of how they operate. If we type 2 + 2 into some system designed for calculation, we usually expect to get 4. If we got 11 instead, we would probably be puzzled. This result could actually be a correct answer in a very unusual circumstance: using a base 3, or ternary, number system, the answer to *2 + 2 is* 11. To figure out what's going on in this unusual circumstance, a person would have to understand something about arithmetic in different bases—binary and hexadecimal being more common ones in computing. That obviously requires additional mathematical knowledge.

But consider a person who truly has no expectation at all of what is going to happen upon entering 2 + 2—no idea at all. The computer might as well output 17, or display the message "Hello, world," or play Rick Astley's music video "Never Gonna Give You Up." If any result seems equally likely, it will not only be impossible to solve somewhat sophisticated puzzles such as discovering the workings of a ternary calculator; it will also be impossible to understand when the computer is doing basic arithmetic, in our ordinary decimal system, and can be used in a straightforward way. Similarly, if we imagine that 4 is always the result of typing anything in or that 4 is a way of displaying the time or that 4 is just a random digit that appears, the system would be as inscrutable to us as a book written in an alphabet we have never seen.

Only because we expect that entering 2 + 2 will result in 4 can we make and test a hypothesis about the system: this is a system for doing arithmetic. We are able to put together three things: our knowledge of arithmetic, our expectation that the system will do arithmetic, and our experience with the system.

Whether you're undertaking early explorations of computing to learn about programming or using computing to explore questions of importance to you, it is extremely helpful, when you are working to develop a program, to have some sort of *expectation* about what will happen when you invoke a function or run a program. This doesn't mean that you always need to know every answer in advance. If you did, why bother using a computer?

But for instance, if you try adding 961907 and 560166, assuming that you are doing integer arithmetic, you should in most cases expect that you will get a number in return. It should be an integer, a number without a decimal point. It should be positive; something's wrong if you tried to add two positive numbers and got a negative result. And if you want to develop a more detailed expectation, you might expect that the number will have seven digits in it, with the first digit being 1, because the first number is almost one million and the second number is a bit less than six hundred thousand. When you see "1522073," you will be able to do some quick checks, such as these, to make sure things are working properly. If you saw, for instance, "961907560166" instead, you'll know that something is amiss.

To digress for a moment about what might have happened in that "961907560166" case: what seems to be amiss in this case is that 961907 and 560166 are being added as strings, not as integers, in the same way that `'work' + 'shop'` is `'workshop'`. The existence of different types for data, such as string and integer, will be covered in the next chapter, and string addition of this sort will also be discussed. For now, the important observation is that even an *approximate* expectation can be of real help when troubleshooting and can allow one to identify when there is a problem in the first place.

If you truly have no expectation about what will happen, your expectation can never be violated and you can never be surprised. This means that you will not be able to tell if your program-in-progress is working properly, if it is indeed doing the useful things that you intended, or if it's doing something else entirely. It also means you won't be able to learn!

Because of the need for such expectations, I ask that you continually take a guess about what is going to happen before you press Shift-Enter. In some cases, the answer may be pretty obvious. Okay, then; take the chance to very quickly confirm that you understand some of the fundamentals of computation. In other cases, you really *need* a computer to figure out the complete result and aren't going to quickly come up with it yourself. Still, you should have some idea about the result—for instance, will the program crash, with some sort of error produced? Will the result be a list of values or a single value? If a single value, will it be larger or smaller than the last result, or will it be the same?

[5.4] Describe the Function

Right now, describe what you believe double() does—not precisely *how* it works internally, but what it does from the perspective you have, your perspective from having called this function a few times, from invoking it on some arguments. Actually write (or type) a statement expressing your current understanding of this function. You can do that right in Jupyter Notebook; there are two simple methods for doing so. The first is just to type # to indicate a comment and then type your statement in the current cell. Alternatively, you can go to the menu and select Cell > Cell Type > Markdown to convert the current cell into one in which you can write text instead of Python code. Then you don't need to include that # at the beginning.

You should write something down or type something out so you can figure out what you are able to describe precisely and what is more difficult to express. Don't proceed until you have committed to some description of this function.

Here is my answer. The operation of double() can be described as follows: "This function takes exactly one sequence as an argument and returns a sequence in which each of the original elements are doubled." I've tried to choose my words carefully, according to the rather precise terminology of programming and Python. I don't expect that a new programmer would come up with exactly the same terms, but this is how I characterize the function.

My description is a more formal way of saying something like "double() goes through each element in a sequence and doubles it," which also gets at the basic idea. The more formal way, however, includes some detail that is important. Having started to think about computation seriously, we need to be discerning about how computation works. It will be very important to see that double() doesn't *modify* the data (which is a sequence) that it is given as an argument. After calling double(x), the variable x is unchanged. Although we will cover variables in values in more detail in the next chapter, see for yourself by assigning a sequence to the variable x and then calling double(x):

```
x = [1, 2, 3]↵
```

```
double(x)↵
```

Now, check to see what the current value of x is:

```
x↵
```

You can see that it didn't change. The informal description, however, suggests that the argument to double() *does* change. The formal description that I gave first is more precise about this.

As mentioned, I haven't actually explained what a variable is, but by trying the previous lines out in Jupyter Notebook, you've gotten at least a little bit of practical experience with the concept. More about variables soon.

Did you make this distinction properly in your written definition of the function, explaining that the original data isn't modified? There is no reason you should have known to do so. I would be surprised if a reader who was truly new to programming had done this. What about the language you used to describe a sequence, each element, and how a sequence is returned? These are also fairly tricky, but all of that should be clear by the end of the next chapter. Hopefully you *did* correctly indicate that the result has "double" each value or twice the value of each element, or that multiplication by two was involved, or, somehow, that an operation of this sort was being done. Did you choose a way of describing this that is clear? Would you revise that part of your description at this point? If so, go ahead and revise it. Really—write down a new, revised description of the function, ideally typing it right into your Jupyter Notebook.

To begin, it can be helpful to simply understand a function as a bundle of computation. After learning more about how computation works, functions can be more clearly distinguished from other things that are also involved with computation. This function "takes . . . an argument," which means that it must be provided some data—not zero pieces of data, and not more than one piece of data, but exactly one argument. We saw already that if we give zero pieces of data (nothing at all between the parentheses of the function), the result is an error. If we provide one sequence, or list (of however many elements), the function seems to work. The single argument that double() accepts is a sequence, which can itself have no elements, one element, or more than one element.

In general, functions *return values* as well as *accepting arguments*, and we'll see that in all of the following chapters. In Jupyter Notebook, whatever values are returned by a function are shown to you when you invoke that function. If you place this code in a text file (we will see how to do this later, in 7.2, Hello World) and run this text file as a Python program from the command line, you will need to print out the result, rather than only having a function that returns a value. Otherwise, the result won't be displayed.

The formal description I offered previously is really all a person would need to carry out this computation manually and is all a programmer would need to implement this function, whether in Python or another language.

The description explains why typing in double([]) succeeds and why it returns []. In this case, the function is given a sequence. Yes, the sequence is empty, but it is still a sequence, just as 0 is still a number. What the function returns—a list that is just as empty—is appropriate. There were no elements in this sequence given, so there was nothing to double. The function, then, did its work trivially; it did nothing. Another way of saying this is "all the elements in the list were doubled—all zero of them!"

Why, then, does double() fail? The error message explains it:

```
TypeError: double() missing 1 required positional argument: 'sequence'
```

We don't have to get into any details of what a TypeError is, right now, to see the problem: double() expects an argument—as our more formal description says, it is missing one required positional argument—and it didn't get one. We can multiply by zero if we want, as in 15 * 0. But we can't just multiply a number without providing a second number, by doing something like 15 *. Recall that 2 + wasn't a syntactically valid expression. When we write a function to work on lists, it's no problem to double the empty list. But it is a problem not to be given a list at all.

We could modify this function, if we wanted, to accept no arguments and return nothing, without producing an error. The error message we see isn't some kind of limitation of the Python programming language. It simply follows from how this code is written. As things stand, double() correctly implements the more formal description of the function. double() "takes exactly one sequence as an argument" according to that description. So if we give it anything else—no arguments, more than one argument, or an argument that isn't a sequence—we shouldn't expect it to work. Try giving double() a number such as 5 all by itself, not surrounded by brackets and thus not part of a list. After you have developed an expectation of what will happen, press Shift-Enter.

Now, see what happens if you give double() two lists as arguments. For instance, you can use the lists [1, 2] and [3, 4]. To hand both of these lists to a function as arguments, you need to place a comma between them. Try to invoke double() on those two lists based on this description of how to do it—and remember that according to the description of this function, you should expect an error to result. Then, proceed to the next paragraph.

If you wish to pass more than one argument to a function, you can do that by putting several values, separated by commas, in between the function's parenthesis. double([1, 2, 3]) will double one list, the list [1, 2, 3]. On the other hand, double([1, 2], [3, 4]) is an attempt to double two lists. Did you come up with that second formulation based on the earlier description? There are several subtleties in play already: lists bundle together several pieces of data, and functions can take no arguments, one argument, or multiple arguments. The function, as written, only takes one argument and won't work with two lists. However, notice that if you give both [1, 2] and [3, 4] as arguments, the resulting error message no longer includes (0 given). The error message does explain what's wrong with that attempt to invoke double(). We'll see how a function that takes multiple arguments works before too long, but hopefully trying out these variations will at least explain why it's important to get parentheses and brackets right. They do indicate different things and are definitely *not* just decorative.

Without even discussing *how* double() works internally, we've gone through a great deal about how it can be *used*. We now know quite a bit about the *interface* to double(). Specifically, this function accepts a single argument, and that argument can be a list, such as a list of integers, such as [1, 2, 3].

Fortunately, no programmer has to write all the code in the world. Programmers can use code that is built-in, code in standard modules, and code provided by others, including people they are collaborating with. Knowing how an interface works empowers a programmer to use other code, including other functions, even if the programmer doesn't write that code or examine that code to learn how it works.

In the remainder of this book, we'll talk more about double() as a way to learn more about programming. The discussion of double() will continue, and the function will be explained entirely and in detail in the next chapter. This quick discussion, with the opportunity for the reader to follow along, is meant to show that it's possible to do some interesting, if very brief, programming within a Python interpreter. An interpreter (Jupyter Notebook, in our case) can be a good environment for figuring out how Python works. It's also meant to lay the groundwork for understanding how computer programs generalize our ability to calculate, allowing us to unfold new capabilities along several dimensions.

As we near the end of this chapter: the end of a chapter is one great point (although perhaps not the only point) at which you should be sure the specific Jupyter Notebook session that you've been using has been given an informative name, for reference. Jupyter Notebook does save your session along the way, but there's no harm in clicking on that floppy disk icon before you log out. If you start Jupyter Notebook in the same directory (i.e., the same folder), you should be able to easily return to your earlier, saved sessions. Or, if you like, you can move the .ipynb file around and open it by starting in whatever directory you have placed it in.

You should be careful about closing and opening Jupyter Notebook files, however. Unless you have pressed Shift-Enter in a particular cell of your Jupyter Notebook, that code hasn't run, despite the fact that you ran it previously. So if you save your work from this session and open it up again, and then go to the last cell at the bottom, you'll find that double() is not defined. If you want to run only the cell where you defined double(), you will need to go to that cell and press Shift-Enter. If you want to rerun everything you did, you can go to the Cell menu and select the Run All option.

[Free Project 5-1] Modifying "Double, Double"

Do this project at least two (2) times, developing different modifications. In the next chapter, there will be exercises that ask you to modify double() in various ways. But without looking ahead to those, figure out some simple but somehow interesting modifications you can make to double() right now. For instance, a simple modification would be creating a new function that prints out triple the value of each number in a list. See if you can do this and if you can think of some other modification beyond that to make, and see if you can implement those. Your changes do not have to be extensive, but they should still reflect good programming practices as you understand them so far. For instance, if you change double() so that it prints out triple the value of each element, shouldn't you change the name of the function, too? You don't have to in order to develop a properly working function, but if you left the name as double(), it would be rather confusing. With the freedom to choose any function name comes the responsibility to choose one that is clear and helpful, not only to others but also to the original programmer who might return to the code months or years later.

Some might protest that this project is being assigned too soon. After all, the discussion of double() has been almost entirely about its interface—how to use it. The only other discussion was about how there is a standard sort of template for function definition. Nothing has been said, so far, about how the code inside the function actually works.

However, in the previous free project, modifying a HTML/JavaScript text generator, there was even less said about the code we were changing. Whatever code you modified was code you selected. Use this opportunity to explore the code in an ad hoc way, making some changes and seeing what results. The method of free modification (done in small steps) works for figuring out JavaScript and can work in Python, too. Give it a try.

Part of the point of double() is that it iterates over a sequence. So as you change the function, don't remove the loop that it uses to go through every element in the sequence.

The only other thing to add is that you should keep it simple. Don't elaborate your new function very much, and definitely don't exceed ten lines. You're not getting paid by the word here. You want to be able to clearly understand how small differences have an effect and how a little bit of additional computation changes the operation of this function.

[5.5] Essential Concepts

[Concept 5-1]   Each Function Has an Interface

Computers don't just offer user interfaces; functions have interfaces, too. In general, they accept arguments and return values, and the specific way in which they do this defines the interface to a function. Be sure you know what an argument is and what it means for a function to return a value. Here is a crucial question you should be able to answer: Can one understand and use an interface of this sort without knowing how the function works internally?

[Concept 5-2]   Code Has Templates

We saw an example of a very standard structure for writing code and abstracted from that the template for defining a function: a function definition always begins with the same keyword, uses parentheses on that first line, has a first line that ends with a colon, and so on. Without looking back at the book, can you write or type it out, with the three blanks or slots, as it was described? If not, review this template before proceeding. Be sure to understand that the indentation is no less important than the punctuation, which is just as important as the three-letter keyword!

[6]   Programming Fundamentals

[6.1]

At this point, I'll describe several types of abstraction. After initial encounters with abstract art, some people have the impression that abstraction is about being nonfigurative or nonrepresentational or making work that doesn't look like the world. Actually, whether it's an abstract painting or the abstract to a scientific paper that one is looking at, abstraction is often better understood as *capturing the essence* of something while not getting bogged down in details.

When writing a program, you may find that in different places and in different contexts, you are doing essentially the same computation ten times. For instance, you might need to take a string that contains a first name, a last name (perhaps a compound one), and possibly one or more middle names, and break it into its component parts. You might need to do this with an email contact over here and with a few other names over there. The first relevant type of abstraction allows the programmer to create one *function* that bundles code together and expresses this essential computation; that function can then be used ten times—or however many times.

You also may find that you have a regular sequence of data to deal with, such that each data point you wish to process should be treated the same way. For instance, you have a bunch of subtotals for services provided and need to compute the tax on them, each in the same way. In this case, instead of (or in addition to) writing a function to deal with a particular number or another single piece of data, a programmer can write a program that goes through a list of this data. That is, *iteration* can be used to generalize a computation over a sequence.

A programmer can also write code that applies to data of different types, so that some code that checks to see if two numbers are equal can be used for people's names (represented as strings) as well. This type of abstraction is called *polymorphism*. And while it isn't as standard to learn early on about this type of abstraction, it is certainly essential for a new programmer to understand *types* and how they help to represent data, along with what sorts of computations can be performed across types and what sorts cannot. Types and polymorphism are closely related.

By considering functions, iteration, and polymorphism/types, we'll begin to see why programming languages, even though they can be used as calculators, are much more than just calculators. These concepts and one other fundamental (the conditional) provide programmers with significant capabilities for working with computation. Through discussing these aspects of abstraction, we'll see how programming can help people get to the essence of important problems and questions—not just in business and science, but also in the arts and humanities.

[6.2]   A Function Abstracts a Bunch of Code

Having mentioned that you could use Python to do your taxes, let's take a quick plunge into world of commerce and taxation, however shallow this world may seem to some artists and humanists. It's very helpful to begin with a concise problem that is somehow grounded in the world. With apologies to those who don't care about the culture of New York City, we'll figure out how much city and state tax there would be for $1,500 of surveillance services provided by a private investigator, working there.

Are gumshoe services taxed in the Big Apple? Yes, they are. The fourth category of goods and services subject to New York State Sales and Use Tax is "detective, cleaning, and maintenance services." And the New York City Finance website, in particular the page www1.nyc.gov/site/finance/taxes/business-nys-sales-tax.page, explains that for this and other items, in 2015, "The City Sales Tax rate is 4.5%, NY State Sales and Use Tax is 4% and the Metropolitan Commuter Transportation District surcharge of 0.375% [results in] a total Sales and Use Tax of 8.875 percent."

The people behind that Web page do this for a living, but let's use Python to check their math and make sure that 8.875 percent is really the right tax rate:

```
4 + 4.5 + 0.375↵
```

If we've typed this in correctly, we'll see that New York City Finance doesn't have its thumb on the scale—the number it gives is consistent with its explanation of the tax—and also we'll see that we can use more than two operands (in this case, with + between them) to add three numbers together. By the way, you don't actually have to type the leading zero in 0.375; just typing .375 will do the same thing. I include leading zeros because I think it makes things clearer for people reading the code.

Now, let's consider how we would figure out a percentage (one percent) of something. There are several possible ways, but let's simply move the decimal point over to the left two places. Moving the decimal place over in 5 by one place gives us 0.5; moving it two gives us 0.05, which is indeed one percent of five. So, we convert 8.875 to 0.08875.

The goal now is to multiply our dollar amount, $1,500, by the number we multiply by to get 8.875 percent, which is 0.08875:

```
1500 * 0.08875↵
```

Let's see what the tax would be on a shorter or simpler job for which only $995.50 is billed:

```
995.50 * 0.08875↵
```

Without doing exhaustive testing, it's easy to do a quick reality check and see that the number computed looks right. It's slightly less than $88.75, which would be the right tax amount for $1,000. (Isn't it? From simple movement of the decimal point?) This tax rate is about 9 percent, so for $100 it should be just less than $9—and that checks out. So it seems that we can use Python (functioning as a calculator) with the number 0.08875 to find out the appropriate tax rate for an arbitrary amount.

However, even with a simple computation like this one, if we were going to do it again and again, it would be a good idea to bundle it up so that we don't have to type in this number every time tax is computed. We will probably also have other things to compute—the cost of a set of prints of incriminating photographs, a surcharge for digging through the trash, and so on. And the tax rate will change once in a while, so even if we do eventually learn to flawlessly type "0.08875", we may find it hard to adjust when a new rate is set.

So, again simply using the Python interpreter, let's define a very simple way to compute the tax at this 8.875 percent rate:

```
def tax(subtotal):
    return subtotal * 0.08875↵
```

As with the function double(), this follows the template for a function definition. Also, recall that the four spaces at the beginning of the second line, which will be automatically inserted for you in Jupyter Notebook, should stay right there. In some other programming languages, indentation is optional, simply for human legibility, and a matter of style and convention. Python, remember, uses indentation to indicate special code blocks and to determine when function definitions, conditional statements, and other constructs start and end. It's part of the syntax, as important as using the word def. In the first line, def indicates a function, just as it did when we defined the double() function earlier. Our function in this case is called tax() and, as with every function, the name is immediately followed by a list of arguments in parentheses. In this case, there is just one argument, subtotal. There don't have to be any arguments, but in all cases, the parentheses are needed. They are part of the template for defining a function. Finally, the line ends in a colon to indicate that there is a special block of code, the body of the function, indented underneath it.

In this case, the special block is only one line long. It contains the keyword return to indicate how to provide the answer—what value the function should send back when it is called or invoked. Let me mention at this point what a *keyword* is. It's a special word that is part of a programming language and is reserved so that it can only be used in particular, formal ways within the language. Many of our choices of what to name things, such as functions, are pretty much free choices. We can call a function double() or tax() if we like, or phoebe() or helicopter(). But we can't call a function return() because return is a keyword, reserved for a special purpose. Don't trust me on this; try it out. Type in the tax() function exactly as it appears previously, but call it return() rather than tax(), and see what happens when you press Shift-Enter. There are not a huge number of keywords that we'll be bumping into accidentally, so while it's good to know what keywords are, we need not be overly concerned about this limitation on naming things.

Back to the matter at hand, calling a function. When we're talking about the use of a function within our code, we would usually refer to such a use as "calling the function" or as "invoking the function," depending upon whether we're feeling telephonic or magical. "Calling" is used often; it just means sending a function some arguments so that the code within the function is executed and a result is returned. The value provided by tax() here is that of the expression subtotal * 0.08875. More on what subtotal means in a bit; for now, let's see how the function works by making use of it (calling it).

Notice that the following line does not have an ↵ at the end. Type it in and *don't* press Shift-Enter:

```
tax(1500)
```

What do you expect this line to do? It's an invocation of the function tax(), just as in the last chapter double([1, 10, 5]) was an invocation of the function double(). In this case, it's not a list but an integer value, 1500, that is being passed into the function tax(). What answer do you expect to result?

You don't have to understand everything about how a function or program works to have an expectation about what it will do. Nor do you have to have computer-like computational abilities to have some expectation about a result. In this case, because we know that the `tax()` function computes tax at the 8.875 percent rate, you can expect that the answer is the same as 1500 * 0.08875.

Having thought about this a bit, go ahead and press Shift-Enter:

↵

Is the answer the one you expected? If it isn't, it is almost certainly because the two lines beginning `def` and `return` weren't typed in properly. To fix them, you can just edit them in the original cell. There aren't many other explanations for what the problem could be. I suppose someone could have possibly played a prank on you by replacing your Python interpreter with a similar-looking program, but that seems unlikely, to say the least.

You should find that typing `tax(____)` and typing `____ * 0.08875` yield exactly the same results when you put the same numbers in the two blanks. Of course, check this yourself with a few values. By doing so, you'll get practice using the Python interpreter to verify your expectations about your code. Or, if there is an error in `tax()`, you'll likely uncover it.

Because `tax()` returns a value, it can be given an argument and used in an expression itself. Here's how to determine the total that a customer should be billed for $1,500 worth of surveillance, which we get by adding the subtotal to the computed tax amount:

```
1500 + tax(1500) ↵
```

If the customer owed $200 from a previous invoice, the total amount owed would be:

```
200 + 1500 + tax(1500) ↵
```

One of the most important ways programmers use abstraction is by placing code in functions. Doing so has several practical and conceptual advantages. Without writing the following program, we'll imagine a billing program that uses `tax()` to compute the tax on private detective services. Let's imagine the completed program uses the function `tax()` in a few places because there are some different contexts in which services are taxed.

Here is a simple-sounding but important advantage of developing a function. The function here is called `tax()`, a name that suggests it is computing how much tax is owed. If we simply wrote `* 0.08875` in the code instead of calling this function, and instead of using the word *tax*, there would be no overt sign that a tax computation is being done. Now, we could add such an indication by writing a comment, which is for people to read and which is ignored by the interpreter:

```
sum = 1500

sum = sum + sum * 0.08875 # Add the tax amount↵
```

This should update the value stored in the variable sum, but you'll of course want to see what the new value is, so:

```
sum↵
```

You've already seen that you can add comments beginning with # in Jupyter Notebook. They can also appear, as here, on the same line as code. Everything from # on is simply a human-readable annotation, having no effect on how the program runs. These comments can be included in any situation, function or no function, to help document our programs. When we develop a function such as tax() or declare a variable such as sum, we are also explicitly given the opportunity to name that function or variable meaningfully. Comments can be very helpful; I believe meaningful *naming* is an even more important practice.

An advantage of a function like this is that it allows the tax computation to be implemented and tested on its own. The formal definition of this function is essentially done in the same way this tax rate is defined in New York City's tax documents and used in manual accounting practices.

Another advantage of placing the very simple expression subtotal * 0.08875 in its own function is that when the tax rate changes, as it surely will at some point, it's possible to update the program by editing it in a single place. If the tax rate is represented in just one place in the code, there is no danger that the old rate might be left in some places and might be updated in others. Avoiding this sort of duplication is an important principle in programming, helpful to explorers as well as professional software engineers: DRY, which stands for *don't repeat yourself*.

It could happen that a more elaborate tax computation—not a single rate—is put in place. For instance, New York City could decide that services below $1,000 will be taxed differently than services that are $1,000 or more. In that case, the update would require something more than just exchanging one number for another. Again, if the tax computation is done in a single place in a function, it would only be necessary to change it in that one place, the function tax().

So for both developing a program in the first place and maintaining it later on, it makes sense to place meaningful computations—lines of code that carry out particular related work—together in a function. And we can get an idea of that even with a function that has only a single, simple line of code in it. When we progress to more involved computations and longer functions (and methods), this should become even more clear.

[6.3] Functions Have Scope

There is a wrinkle to the power of functions: the ability to bundle code into them. A function creates a special universe of its own with its own set of variables—different from the variables in the outer universe, where the function is being called. Try the following:

```
def set_a():

    a = 10

    print('The value of a:', a)↵
```

That defines a function in the computing sense. There's no return in this one, though, so this function doesn't provide an answer. This one just displays an informative message indicating that a has been set to 10. Try it:

```
set_a()↵
```

Now, to get really trippy, try this:

```
a = 20

print('Initially, the value of a:', a)

set_a()

print('Finally, the value of a:', a)↵
```

Unless you completely understood my brief comments about a function having its own special universe, this result should look curious to you—at least, the first time you see it. What's happening here is an effect of *scope*. The variable a that is inside the function set_a() is bound to that function; it doesn't matter if there happens to be another variable outside that function named a that has some value. In addition, the a that exists outside of that function isn't affected by what goes on inside. Initially, that a in the outer universe is 20; within the function, the function's own special-universe a is set to 10; and after the function is invoked, the "outer" a is still 20. What's happening with variables *within* the function is actually separate from how variables are set, and what values they have, in the outer universe, the level where that function is being called. What happens in set_a() stays in set_a().

Think for a bit about what would happen if we didn't have scope. In a large project, my function (imagine a real function that is doing some work, not this simple set_a() example) might be called in a hundred different contexts (or even more). If scope didn't exist, none of the variable names used in those outer contexts could also be used within the function. I would need to write a function that didn't use the variable a or num or whatever else occurs in all of those other contexts. Perhaps I would end up using some cryptic variable name such as plmoknijbuhvygc to avoid repeating the name of some other variables being used in the context. What if my function calls itself? As we'll see later, this is possible. Scope takes a little bit of work to understand. But without scope, programming would be much messier.

Having all variables remain "active" inside a function would do a lot to wipe out the benefits of abstraction and code bundling. It makes much more sense to have scope. If a programmer wants to pass in some values that the function can work on, the programmer can do so explicitly in the *arguments* to the function, not in cryptic and possibly accidental ways by having a variable on the outside somehow entwine with one on the inside. But this does mean that programmers need to consider the calling of functions and nesting of code; a in one part of the program may not be a everywhere.

Here's one more example of a function being defined and invoked, this one a bit more complex:

```
def scoped(first, second):

    third = second + second - first

    return third

first = 10

second = 11

third = 12

scoped(2, 4)↵

first↵

second↵

third↵
```

When first, second, and third are used at the top of the snippet of code, in scoped(), those names apply only within the scope of that function. The variables first, second, and third that are defined right after the function definition apply in that "outside

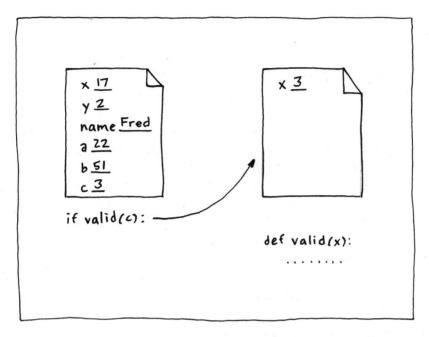

[Figure 6-1]
A new piece of scratch paper is taken out (conceptually) when a function such as valid() is invoked. The only thing on that paper initially is whatever arguments the function has, with whatever values were passed in to the function. Here, the function call is valid(c) and c has the value 3. That means that when this function, defined as having the single argument x, begins its work, it will have only x (associated with the value 3) on its scratch paper.

universe" and could have been anything, including a, b, and c. So the values given to them before scoped() is called remain the same after the function is called.

The function scoped() has two arguments, which are called first and second. Those two arguments effectively become variables within the function. They are limited to that function's scope. When execution of the function ends with return third and the corresponding value is sent back, those local first and second variables are no more. The scratch paper on which computation was being done is thrown away, and the program continues executing at the higher level, where different variables, first and second, are defined.

Up at the top, in the first three lines, first, second, and third could have been called la, lala, and lalala and the function would work in exactly the same way. Prove it to yourself by making the change, substituting those three names for first, second, and third in the function definition, in those first three lines only. You do need to change both occurrences of first, all three occurrences of second, and both occurrences of third.

I used the word *universe* because, while it sounds a bit far out, it is an existing word in common use and a concept that can help some people to understand scope. Another way to explain that a certain variable, such as a, i, or num, can exist in different universes of these sorts is to say that these variables are situated in different contexts. And it's also possible to think of a function as getting out scratch paper and doing its work with whatever variable names it likes, then discarding the scratch paper when it finishes its task and returns a value. The *arguments* to the function hold the values that the function needs to do its work, and once it gets an answer it sends that back using the *return value*. What happens in between is just temporary computation taking place on the function's scratch paper.

The benefits of scope, like the general benefits of bundling your code into functions, become clear when one starts writing longer programs. If you're just writing a few lines of code to try out some ideas and quickly explore a concept, having these multiple universes—or multiple contexts—or a scratch pad—may not be so advantageous. But when you start to scale up and write even slightly more elaborate programs, knowing about functions and scope will be essential.

[6.4]   Iteration (Looping) Abstracts along Sequences

One of the simple but important capabilities of computers is that, after computation is bundled up in some way (such as in a function), it can be applied many, many times. For a computer, it's about as easy to do a computation one time or a million times. If you host a popular forum on the Web and notice, while reading along, that someone has used a lot of exclamation points in a post, you might decide to investigate the use of punctuation by members of the forum. Your computer can go through tens of thousands of posts and count the exclamation points in each one in about as much time as it would take you to examine a single post. You could then see if the use of exclamation points is more highly correlated with particular threads, times of year, users, or something else. And, of course, it would not be much more difficult to look at more sophisticated statistics of posts that go beyond counting occurrences of single characters: a simple statistic such as average word length could provide a crude estimate of the reading level of posts. More elaborate techniques such as topic and language modeling, and, for instance, sentiment analysis, are not beyond the reach of researchers who have a bit of background in programming and a willingness to explore.

But to get started on a process such as this one, it's important to be able to write a program that can *iterate*—run through sequences of data, doing the same computation on each one. Back in the day, computer programs were encoded on punched paper tape, and to run a program, this tape ran through machinery. Iteration could be achieved by actually making the paper tape into a literal loop, so that the computer would keep processing it again and again. The general concept of looping (or iteration), then, isn't hard to grasp. After understanding more detailed aspects of this concept, it will be easy for a programmer to generalize computation to lists and other sequences.

To begin, let's look at a very simple type of iteration which has the effect of displaying every element in a list on screen. This is commonly called *printing*, and most people are quite comfortable with the term being used in this way, although Gutenberg would probably be quite surprised at this particular usage.

We can define a list like so—using the letter *l*, not the numeral *1*:

```
l = [7, 4, 2, 6]
```

This looks a lot like the line x = [10, 20, 30]—a line we saw and input into Jupyter Notebook in the previous chapter. Let's be clear, at this point, about what this line of code means. We are not asserting that l *equals* [7, 4, 2, 6] here, nor are we asking the computer to tell us if l is equal to that list. Even though we are using the symbol that is well-known as the equal-sign, we are actually not dealing with equality. In Python, it is necessary to use two equal-signs (==) when we want to check whether one value or variable is equal to another.

This line, instead, is performing an *assignment*. It declares that the variable named l will be assigned the value [7, 4, 2, 6]. Or, to phrase that differently, it means that the list of numbers [7, 4, 2, 6] will be given the label l so that we can refer to it by that label.

Type this in to iterate over (or iterate through) the list l like so:

```
for num in l:
    print(num)
```

The for statement indicates iteration and declares that the program should do the same thing "for" each element of the list, which in this case is named l. Each individual element is, as we go through the list, given the label num, so that initially num is 7, then num is 4, then num is 2, and finally num is 6. And at each step, a simple operation is performed: the value of num is displayed on the screen using print().

This is a simpler process than seen in the function double(), but notice that it does *not* return a value that can be used in further computation. It just displays data for a person to read—perhaps in an informative fashion, but not in a way that can be built upon. The functions we write typically *return* the values we're trying to compute. However, we have the option of using `print()` to provide us with information along the way, as we develop and debug our functions.

There is a kind of pattern for iteration, one that looks like this:

```
for ____ in _____:

    ____
```

In the first blank is the variable that will be used to hold each element. It can be called almost anything; num is used in the preceding example, but it might be i or `element` or anything that isn't used as a Python keyword.

In the second blank is the sequence to iterate through. I'm using *sequence* without giving an official definition. The term actually has an official definition in Python, but our folk understanding works well enough for now. A sequence can be what a function generates or returns. In general, we might use a sequence that is something other than a list. In our discussion so far, what we have is the list labeled 1, and it's good enough to understand a sequence as some elements in order.

After that second blank is an important part of the template, although it's a single character, :, a colon. This is necessary to indicate that everything that is indented underneath is part of this iteration and included within this for loop.

In the third blank—and possibly occupying more than one line of code, all within this same level of indentation—is whatever we want done during each iteration. This is the code that will be run "for" each element of the list.

When you're studying how to iterate through a sequence, you'll see *for* and *in* very often because they are fixed parts of the iteration template and, indeed, are keywords. The name of the variable that holds each element, and the name of the list, is free for the programmer to determine, as is the code that runs each time.

Having tried the first version, try this variation:

```
for num in 1:

    print(num * 2)↵
```

Now you can see that, in addition to just enumerating the list, and displaying each element, there is the potential to *compute* at every step. Nothing very exciting has happened yet, but it should now be evident that iteration allows true computation to take place; it's more than simply a way of displaying the elements. Again, although we've used `print()` here to provide an expedient display, we will generally write functions to return values instead—as in the next example.

Let's take a straightforward (if not completely exciting) operation that we can do on any sequence of numbers: we can compute the average, or, to be very specific (as there is more than one type of average), what is called the *mean*. That's what is most commonly called the average—in issuing grades, for instance, or determining the average number of runs a baseball player has batted in per game. It's the number that is the sum of all the data points (7 + 4 + 2 + 6 in the list 1, for example) divided by the number of data points (there are four in the list 1). For two points, it's simply the sum of those two numbers divided by two. For three points, it's the sum of those three numbers divided by three. And so on for any number of points.

Here's a way to compute the mean of any sequence and return it. This code is a bit longer than the very short snippets seen so far, but you can go back up and edit it in Jupyter Notebook if you make a mistake. First, we're going to type in this function and leave out an important part of it. We'll see what error crops up, and then we'll fix the function.

```
def mean(sequence):
    for element in sequence:
        total = total + element
    return total / len(sequence)↵
```

You'll need to remove the indent to one level (or four spaces) to enter that last line. This represents one code block (the iteration) ending and the next line beginning in the previous code block (the function definition). You can just tap the Backspace key, or, on a Mac, the Delete key, a single time to go back one level of indentation.

Division (indicated with the / operator) was mentioned earlier but is being used in this book for the first time here. There's nothing particularly special about the way division works in Python 3.

WARNING! If for some reason you are using Python 2 instead of Python 3, which you should not be doing when you use this book, division is actually defined differently! Use Python 3 with *Exploratory Programming for the Arts and Humanities, Second Edition*.

This code is, again, defining a *function*. It has a for loop that iterates through every element of the sequence, which is passed in as an argument and is meant to accumulate the total. Then it returns that total divided by the number of elements, which is the mean.

However, it won't work right away. Run it—or to be precise, call it on some sequence of numbers—to see why not.

The problem is that this code instructs the computer to use the value `total` (on the right-hand side of `total = total + element`) before any value has been assigned to it. It's like asking the computer for the answer to `temperature - 80` before you have given any value to `temperature`. Go ahead and type `temperature - 80` into Jupyter Notebook, press Shift-Enter, and see what happens. The way to fix this problem in our particular case is by adding an extra line, as follows:

```
def mean(sequence):

    total = 0

    for element in sequence:

        total = total + element

    return total / len(sequence)
```

The first thing done in the new version of the function is in the line `total = 0`. That line is *initializing* the variable `total`. It assigns the value 0 to `total`, which is indeed the value that should be in there before anything has been accumulated.

Initialization of this sort makes good conceptual sense. If you're counting up how many books you have in some row of boxes and you're keeping your answer on a notepad, before you open that first box, your *initial* answer is 0. You haven't seen any books yet. Of course, if you were doing this activity yourself, you probably wouldn't write down a zero on your notepad to begin with. You would just open the first box. But in Python (as with other programming languages), we have to say that we want to use a variable named `total` and we have to give it some explicit value to begin with. Otherwise, we can't increase the value of total by assigning it using the line `total = total + element`.

After we've added this initialization, as the for loop iterates through each element in the sequence, the value of each element is added to `total`. Finally, the value that's returned is the accumulated `total` divided by the length of the sequence.

Try this function out on a short list in which the mean is obvious—for instance, one in which every element has the same value:

```
short = [5, 5, 5, 5]
```

```
mean(short)
```

The mean should certainly be 5. It almost is, but notice that the answer has a decimal point followed by a zero. At the final step, when division occurs, the resulting value is not an integer, but of a different type. Consider this a preview: types are discussed in the very next section, 6.5, Polymorphism Abstracts across Types.

Next, try defining a more interesting, but still fairly short, list, and see what the result is. For instance, you can tell that [-5, 0, 5] should have 0 (in this case, represented as 0.0) as its mean.

You should be able to prove to your own satisfaction that you have a function that is completely general to a sequence of any length. Let's be clear about what that means: whether you have a list with four numbers in it or ten thousand, this function will find the mean. I actually defined a list of ten thousand elements in a Python interpreter to make sure that this would work and to see mean() functioning properly on this list. Indeed, everything worked. I was sure this would work already, but I wanted to be able to truthfully write that I had done this, to make it clear that you *can* prove these things to yourself. This means that, just by using a for loop, the computer is able to amplify your powers of analysis thousands and even millions of times. In this simple case, we've done a numerical operation, but text and other media can also be treated in this way, as we will see soon.

[6.5]   Polymorphism Abstracts across Types

We'll investigate the concept of types by examining three different values and by checking to see what type each of these values is. Once we get a handle on types, we'll be able to see how a mouthful of a concept (polymorphism) allows for abstraction across different types.

A value is a specific piece of data. Values can be stored in variables, which can be understood as slots or labels for different sorts of values. So the variable x, for instance, can be used to label any value:

```
x = 5↵
```

And we can check to see what type of value the variable x holds by using the built-in type() function:

```
type(x)↵
```

Notice that we get the same answer if we just check the type of 5 directly:

```
type(5)↵
```

This type, *int*, is what Python uses to represent integers—numbers without a decimal point. Let's check what type 'hello world' is:

```
x = 'hello world'
```

```
type(x)↵
```

Notice that the type associated with the variable x has changed, as it should have. This *str* is a string, a sequence of characters, which is what 'hello world' is. Before continuing, check to see what the type of 'hello world' is by directly checking that value, just as you checked what type(5) is.

Now, use type() to see the type of the list [1, 2, 3]. This function tells us that the type is *list*. We could actually be more specific when describing this list: it is a list of integers.

As you program, writing functions and using them, you should have a sense for what type of argument, or arguments, a function accepts and what type of result it will return. It's part of understanding what computation you have specified.

A handy feature of Python is that certain functions, and certain operators, work on several different types. For instance, given two strings, a and b, it's possible to join these strings together in two ways, with a at the beginning or with b at the beginning. The operation of putting two strings together is called *concatenation* and is done in Python as addition, using +. So as you think about what you expect will happen, try typing:

```
'hello' + 'world'
```

And, after guessing what will result and pressing Shift-Enter, try:

```
'knock' + 'knock'
```

Again, wait to make a guess before pressing Shift-Enter so that you can see if you get the result that you expect.

You can also assign the strings you are using to variables and then concatenate the variables:

```
a = 'hello'

b = 'world'

a + b⏎
```

Let's also use another built-in function, len(), which provides the length of its argument.

```
len('hello world')⏎

len([1, 2, 3])⏎
```

You can see quickly that len() works on both strings and lists. For strings, it returns how many characters are in the string. For lists, it returns how many elements the list has. Both of these are fairly easy to understand as being the length of these sequences. Try this and, before you press Shift-Enter, take a guess about what the result will be:

```
len(['hello', 'world'])
```

What exactly is that argument that we handed to len()? It's a list . . . that has a certain number of elements . . . each of which is a sequence, specifically a string . . . and each of which has a certain number of elements. Okay, make your guess and press Shift-Enter. When we check the length of some sequence, it doesn't matter if it's a sequence of sequences or of some other types. This built-in function just returns how many elements the sequence (i.e., the outermost one) has.

You know too that + is used for addition and that Python understands 2 + 2 as being 4. Python also returns the result 'helloworld' when asked to compute 'hello' + 'world'. Thus, + also means concatenation, putting strings together, when the arguments are strings. But not everything has a length and not everything can be added together. Try these to see some types that don't work in these ways, by typing first:

```
len(5)↵
```

and then:

```
2 + 'hello' ↵
```

In the first case, Python doesn't understand integers as having a length. It's not that the concept is completely senseless; an integer does have a particular number of digits, and some numbers could be considered longer than others. It's just that this aspect of integers—their length—isn't represented in Python's model of what an integer is. Similarly, we can imagine adding the number 2 to the string 'hello'—and we could come up with a rule that would allow us to put integers and strings together in some consistent way. In fact, these additions can be done in other programming languages. But Python doesn't provide for an int and a str to be added together.

These TypeError messages are produced because trying to take the length of an integer and trying to add an integer to a string are probably mistakes. Code that does these things is likely to be wrong. If we really want to add the number 2 to 'hello' and get '2hello' as a result, or if we really want to see how many digits are in the number 5, how long it is, we can convert those integers to strings before attempting these operations, as in:

```
len(str(5))↵
```

and:

```
str(2) + 'hello'↵
```

Try these out! This tells Python to treat these integers as strings and tell us how many characters are in the string (in the first case) and that the lexical unit '2' should be concatenated, string-like, to `'hello'`. If we want to treat 5 and 2 like strings, we can explicitly make them into strings. Converting between types is called *casting* and is an aspect of programming that is typically encountered early on and used fairly often.

When a single arithmetic operator, function, or method is able to work on several different types of data, such as integers, strings, and lists, this is an example of *polymorphism*. If you use a programming language such as Python that implements polymorphism for things such as equality and arithmetic, you will find that programs you write with one data type in mind will sometimes work perfectly well on other types, and in other cases they can be adapted easily to work on different types of data.

Polymorphism is also something that can be designed and implemented in particular cases. For now, instead of trying to go into detail, I'll just mention that a programmer can "manually" implement functions that accept arguments of different types (both strings and integers, for instance) and provide the right answers in either case.

[6.6]   Revisiting "Double, Double"

It may be clearer, at this point, why `double()` was introduced as the first snippet of code in this book. This function, in only five lines, exhibits all three types of abstraction discussed so far.

In it, code is bundled together in a function. It didn't *have* to be this way. We can write code that isn't in a function and that does the same thing. Try it out:

```
volume = [4.0, 2.0, 3.0, 5.5]

result = []

for element in volume:

    result = result + [element * 2]

result↵
```

That code takes a list of volume levels and doubles it, using the same technique that is bundled into `double()`. The answer ends up in the variable `result`. (Notice that after doubling, my volume level actually goes to 11!) This is very much like an example of iteration given previously.

Because we've taken the essential process of `double()` and pulled it out of the function, let's go one step further to try to understand what's happening in the line within the loop. To see what's happening, we'll unroll this loop and step through it:

```
result = []

result↵
```

This shouldn't be a surprise. The variable was assigned the value of the empty list, and it has that value.

Now perform the first pass through the list, where the first (doubled) element gets added:

```
result = result + [4.0]
result↵
```

Now the second element:

```
result = result + [2.0]
result↵
```

Can you see how the list is being built up? At each step, a tiny (one-element) list is being added (concatenated) to the current `result`. Go ahead and finish adding the final two elements in this way, to see that you get the right answer.

Now, why are there those brackets around each value? What would happen if we did the following? Make a *guess* about the outcome before you press Shift-Enter!

```
result = []
result = result + 4.0
```

Is this addition going to work? What are the two things being added? Or, at a higher level, what are the *types* of the two things being added? One is `[]`, the empty list . . . and the other is 4.0, a *float* (i.e., a number with a decimal point). Can we add these together? After you guess yes or no, press Shift-Enter and you'll know for sure.

To build up the list that is our final answer, our "result," we can't add 4.0, then 2.0, then 3.0, then 5.5. We have to add (or concatenate) a list with a list. So, starting with `[]`, we add `[4.0]`, then `[2.0]`, then `[3.0]`, then `[5.5]`.

To learn about what's going on, it's useful to pop the hood and see what's happening step by step. But the general problem with writing offhand code, not contained in a function, is that if we want to double the volume levels again, or double anything else, we have to enter those two lines beginning with `for element` again or copy and paste them or the like. Whether we retype them or select and copy, we have the potential to make mistakes, and we will definitely make our overall program longer and harder to work with. We wouldn't be following the good advice of DRY (don't repeat yourself). With the way `double()` is written, on the other hand, that code is usefully bundled and can be easily reused.

`double()` also does iteration, going along and doubling each of the elements in a sequence, whether there are zero or several thousand. This function does it using a loop—specifically, a for loop.

In addition, `double()` implements polymorphism as well, although there are no special signs of that in the code. Because this function computes * 2 for each element of a sequence, it can work on any element that can be multiplied. The volume level example earlier shows that the code in `double()` can double not just integers, as we saw earlier, but also floating-point numbers. (Try doubling `volume` using `double()` to see that the function itself works the same way.) This function can also double strings:

```
words = ['hello', 'world']↵

double(words)↵
```

This versatility comes for free, thanks to the use of multiplication (which can work on several different data types) within the for loop. Notice that we can even have a list with different types of elements, and as long as each one can be multiplied, `double()` can take the list as an argument:

```
countdown = [3, 2, 1, 'contact']↵

double(countdown)↵
```

And, finally, recall that strings and lists are both sequences, and the way `double()` was written certainly suggests that it will work on sequences. So instead of giving this function a list as an argument, give it a string—and *wait* before pressing Shift-Enter:

```
double('abstraction')
```

What do you expect will happen? Formulate a clear guess of some sort. Then, find out:

```
↵
```

[6.7] Testing Equality, `True` and `False`

Try this expression in your Jupyter Notebook:

```
2 + 2 == 4↵
```

I mentioned (briefly) that `==` is different than `=` in Python. A single equal-sign is what is used for assignment, giving a variable a value. Two equal-signs are used to test if the left-hand side and the right-hand side are equal. When I first asked you to work with Jupyter Notebook, in 4.2, Calculations, I asked that you see what 2 + 2 evaluates to: the answer is 4, of course, and Python gets this answer right. Now, the line you just tried asks if the left-hand side and right-hand side evaluate to the same thing. In this case, unsurprisingly, yes. So Python gives us the answer `True`.

Now try checking to see if 2 + 2 evaluates to the same thing as 3. You know the answer, but see what Python tells you when you ask the question. Try once more, and see if 2 + 2 is equal to 3 + 1; then, check to see if 2 + 2 is equal to 2 + 1.

What Python is providing as an answer to your question is not a string, like `'hello'`. You may have guessed that because the value does not have quotation marks around it. It is data of a different type, and this type is called Boolean. This data type gets to be capitalized, unlike integer, string, and the other types we'll encounter, because it is named after a person, George Boole.

To glance ahead briefly, it's not only a test for equality that can result in a Boolean value of `True` or `False`. A test for inequality (e.g., is the left-hand side less than the right-hand side?) will also result in this type of value. For now, let's start with == and see what we can do with that.

Checking equality is most useful when there's at least one variable involved. For instance, try this (two-step) equality test to see that variables can be included in our equality tests:

```
greeting = 'hello'↵
```

This part assigns a value to the variable greeting; it doesn't conduct a test. So nothing is returned; we just have this assignment done after Shift-Enter has been pressed.

```
greeting == 'hello'↵
```

This part is the test, using the == operator. Let's try a different test, which we hope will give us a different result:

```
greeting == 'hi'↵
```

Try doing a few other variable assignments and equality tests, using some data of different types. Integers, for instance, or sequences, if you like.

[6.8]   The Conditional, `if`

Type this simple function into the Python interpreter:

```
def secret(word):
    if word == 'please':
        return 'Yes!'↵
```

Notice that this includes that test for equality using ==, which is completely different from assigning a value to a variable using =. Now, try it out:

```
secret('hello')↵
```

```
secret('world')↵
```

```
secret('please')↵
```

The first two tries do nothing. They don't cause errors; they simply result in the function concluding with no output. The third call returns the value 'Yes!' to signal that the correct word was input. This shows the essence of the *conditional* statement, also called the *if* statement. It causes certain code to run only *if* the *condition* is true. (In Python, as was just discussed in the last section, the truth or falsity of a condition is represented by the special values True and False.) In this case, the word *please* has to be given as an argument to the function, and the word has to be passed in exactly as stated: entirely in lowercase and without spaces or anything else on either side. Try typing in secret(' please') and secret('Please') to see for yourself that these won't work.

Here's another similar function that uses the conditional. It checks to see if the argument is the string 'yes' (just as secret() checks to see if it is 'please') and returns the Boolean value True if it is. Then, it checks to see if the argument is 'no' and returns the only other Boolean value, False, if it is. If the argument is anything else, it does not return a value at all:

```
def yesno(word):

    if word == 'yes':

        return True

    if word == 'no':

        return False↵
```

Try out yesno() and see if it provides the appropriate results—for instance, by typing:

```
yesno('haha')↵
```

```
yesno('no')↵
```

```
yesno('yes')↵
```

```
yesno('Yes')↵
```

This function could be improved; you might want it to recognize 'y' and 'n' as well, and it should probably ignore the case of the string. But perhaps you can already imagine how it might be used in a program that prompts the user for a yes/no reply. If the code to check for an affirmative or negative reply was bundled in a function like this, the same code could easily be used in several places in the program. And improvements that were made (to ignore the case of the reply, for instance) could be made in one place and would apply everywhere. If you wanted to translate this function to work in a different language, that too could be done once in a single place.

Now, let's go back to the original secret() and modify it so that it returns 'Yes!' if the word 'please' has been entered and otherwise returns 'No':

```
if word == 'please':

    return 'Yes!'

else:

    return 'No.'
```

Take a look at what this code does. It uses a new keyword, else, to indicate some (indented) code that should run "otherwise"—when the if condition doesn't hold. What if we want secret() to return True or False instead of those two strings?

```
def secret(word):

    if word == 'please':

        return True

    else:

        return False
```

This code works, but it turns out there is a very nice way to *refactor* it. When we revise code in order to refactor, we aim to keep the function of the code exactly the same, but to make it easier for people (including the original programmer) to read, understand, modify, and further develop. We want True when word has the value 'please' and False otherwise, so we could write this new version of secret() as follows:

```
def secret(word):

    return (word == 'please')
```

In this particular case, the whole conditional expression is superfluous. We really just want to return whether or not the value of word is the string 'please'. So we can simply return the result of that equality test.

This doesn't mean that the conditional is useless. We used it in yesno() a bit earlier. It will have plenty of other uses, as we'll see. But sometimes there are ways to simplify, and a beloved construct such as the conditional turns out to be unnecessary.

[6.9]   Division, a Special Error, and Types

To complete the discussion of basic arithmetic, just a few more brief comments on division. One special thing about division is that unlike addition, subtraction, and multiplication, it is not defined for every set of operands; you can add, subtract, and multiply *any* number with any number of the same type, but that isn't true for division. If we're talking about the integers (numbers without a decimal point, including positive numbers, negative numbers, and zero), A + B has an answer for every A and B. The same is true for A – B and A × B. The reason that this isn't true for A / B is because division by zero is undefined. There is simply no answer to A / 0 for any value A, even if A itself is also 0. No calculator, math teacher, or accountant can find one for you, and the Python interpreter can't either. Try dividing by zero in Jupyter Notebook and see what error is produced.

So, consider the consequences. If you write a program that generates a random number B, possibly 0, or that accepts input from the user and allows the user to define the value B, possibly as 0, then if that program continues and tries to calculate A / B, it can crash. When you're writing short, ad hoc, exploratory programs for your own use, this might not matter; if it happens rarely, you could just run your program again, and, whatever. But generally, it does matter. Fortunately, there's an easy fix for this: don't ever allow your program to attempt to divide by zero! It's an operation that makes no mathematical sense, and a program that attempts this operation isn't constructed properly. Later, I'll discuss practical ways to prevent a calculation from happening in special cases such as this one.

Now that we know a little about *polymorphism* (the way the same operation can apply across types), here's another wrinkle to division. Notice that strings can be multiplied by integers:

```
'hi' * 2↵
```

But they cannot be divided by integers:

```
'hihi' / 2↵
```

That may seem a little unusual, but there are gaps of this sort when operations are made to work across types. The analogy to arithmetic on numbers simply isn't complete in this case. It would be tricky to make string division work, and while it could possibly be defined in a consistent way, it's not clear that it would be worth the effort.

[Exercise 6-1]   Half of Each
Write a function half() that is very much like double(), but instead of returning a new sequence with double each element in the original one, it returns a new sequence

with *half* of each element in the original one. That is, if the first element in your initial sequence is 6.6, the first element in your answer will be 3.3. Your function just needs to work on sequences of numbers. You can use the double() code and very slightly modify it. Don't strain yourself! There are two "easy" ways to do this that I can think of that both involve very slight modifications of double(); see if you can find more than one straightforward way to accomplish this task. The two methods use different operators.

[Exercise 6-2]   Exclamation

Write a function exclaim() that is very much like double(), but instead of returning a new sequence with double each element in the original one, it returns a new sequence with an exclamation point added to the end of each element in the original one. That is, exclaim(['hello', 'world']) will return ['hello!', 'world!']. Again, you can use the double() code and modify it. To begin, get your function working so that when you give exclaim a list of strings, it works properly. Then, as a second step, see if you can use *casting* as described in 6.5, Polymorphism Abstracts across Types so that whatever the elements are, the function will work. For instance, after this second step, exclaim([1, 2.2, True]) will return ['1!', '2.2!', 'True!'].

[Exercise 6-3]   Emptiness

Write a function has_no_elements() that accepts a sequence and returns a Boolean. The function returns True if the sequence is empty—that is, if it has no elements—and False otherwise.

[Exercise 6-4]   Sum Three

Write a function sum_three() that accepts three values—*not a sequence*, but three arguments—and returns the sum of those three values. At the end of 6.3, Functions Have Scope, there is an example of a function, scoped(), that takes two arguments. Given this, you have already seen and typed in functions that accept zero, one, and two arguments, so now generalize to three.

[Exercise 6-5]   Ten Times Each

Write a function ten_times_each() that is very much like double(), but instead of returning a new sequence with double each element in the original one, it returns a new sequence with *ten times* of each element in the original one. You can use the double() code and slightly modify it. There's no trick here! It's an exercise in not over-thinking it: something that should take a few seconds. This is just to get practice and cement your understanding of functions and of double().

[Exercise 6-6]   Positive Numbers
Here is a two-step exercise that asks you to develop functions that combine iteration and the conditional. In this exercise, you need to accumulate a value in two different ways.

First, define positive(), a function that takes a sequence as an argument and counts up and returns how many positive numbers (numbers greater than 0) are in that sequence. You will want to notice that not only can you test equality using ==, you can also test for inequality using < (less than) and > (greater than).

Having done that, start with positive() and modify it to create pluses(). This function should compute and return the sum of all positive numbers—ignoring any numbers in the sequence that are negative or zero.

[Free Project 6-1]   Further Modifications to "Double, Double"
Do this project at least two (2) times, developing different modifications. Now that we've seen more about how double() works and understand some of the fundamentals of computation related to that, modify the function again to make it do something interesting (according to you) when it is given a list of strings. See if you can keep many of the aspects of double() while changing at least one aspect significantly. Your function does not have to work on any other sort of list. You also don't need to skip ahead to figure out any specifics of how strings work. Just use what you've learned already. Of course, don't exactly duplicate any of the previous exercises, all of which you are supposed to complete in addition to the free project.

[6.10]   Essential Concepts

[Concept 6-1]   Variables Hold Values
A variable can be thought of as a "bucket" or container into which different values can be placed. Or, if you like, you can think of it as a "tag" or label that is given to a particular value. Variables can vary, in that they can contain or label many different values, depending upon what is assigned to them using the = operator.

[Concept 6-2]   Functions Bundle Code
A powerful way to write code is to bundle it together in functions. If you're really only going to use some code once, there's no need to do this. In general, however, programmers will want to use the same bit of code again and again, not only to compute a tax amount but also to process every word or line in a text, every pixel in an image, or every text or image in a corpus. Whenever you complete code that you *might* use or deliver later, bundle the essential computations up into one or more functions.

[Concept 6-3]   Functions Have Scope

Functions define their own universes, or contexts. They are their own pieces of scratch paper. They have values handed in via arguments. Inside the function, any variable names can be used, regardless of whether there is a variable outside with some value assigned to it. When the function is done, it can return a value; the scratch paper is always thrown away.

[Concept 6-4]   Iteration Allows Repeated Computation

Whatever can be done to an element of a sequence (such as a list) can be done to every element, using the for statement. Be able to take any function that works on one element and apply it to every element in a list. You really must be able to generalize computation in this way or it won't be possible to progress much further. Finally, be able to accumulate a result, in the way that double() does by building a list, in the way positive() does by counting, and in the way pluses() does by accumulating a total.

[Concept 6-5]   Types Distinguish Data

Understand that there are different types of data and that you can determine the type of a particular value held in a variable. See that different computations apply to different types—some only work on sequences such as lists, some only work on numbers, and so on. Understand that some functions and arithmetic operators are polymorphic and apply to different types. Have a good sense of the essential differences between integers, floating-point numbers, strings, Boolean values, and sequences.

[Concept 6-6]   Equality Can Be Tested

Just as Python will evaluate arithmetic expressions such as 2 + 2 for you, it will also evaluate *equations* such as 2 == 2, as this one would be written in Python. Understand that the result will be one of the two Boolean values and that you can check to see if a variable has a particular value using the equality operator, ==. In doing the exercises, you should have figured out that inequality (whether one value is greater than or less than another) can be tested, too.

[Concept 6-7]   The Conditional Selects between Options

The if statement in Python allows for one computation to be done in one case and another computation (or no computation at all) to be done otherwise. There are several formulations of the conditional, but it's important to understand the essential ability of the if statement to determine whether or not a condition holds and to apply computation if it does.

[7]   Standard Starting Points

[7.1]

Computer programs can be rhetorical, sometimes quite obviously so. Some rhetorical aspects become particularly clear when considering the particular programs that are used to introduce programming and computation. A *starter program* can suggest a great deal about computing. In this chapter, three conventional starter programs are presented for the reader to type in and run. The chapter explains how each program works and uses these programs (as originally intended) to introduce programming. In addition, these programs are discussed critically and in the historical context of computing. The discussion covers the rhetoric, or poetics, of these programs, as well as some of what they highlight and what they hide about programming.

[7.2]   Hello World

Making the computer "speak," and say hello, is often the first task done when learning to program in an imperative programming language. (There are other idioms of programming, but the imperative one considers, at a very high level, that programming is giving instructions to the computer.) The "Hello World" program, on some levels, frames the computer as obedient, autonomous, and able to communicate verbally, albeit in a very simple way. It is also a simple, short program that, although it isn't a stunning display of computer power, is sufficient to show that a programming environment has been set up and is running properly. This is perhaps the most essential function of the "Hello World" program—ensuring that one's programming environment works.

By now, it should be clear to you that Python and Jupyter Notebook work. Still, let's see "Hello World" in operation. Type in:

```
print('Hello World')↵
```

Very simple, and very good. Now, let's take the opportunity provided by this very simple program to figure out another way to run Python code. Python programs are text files and can be composed and edited in any true text editor. Our next step is to create a text file that simply has that single line in it.

You set up a text editor (if you didn't have one handy already) before starting on chapter 3, "Modifying a Program." Open up a new file in your text editor and type in the following:

```
print('Hello World')
```

Save this file as hi.py; the *.py* is a file extension, one that indicates this is a Python program. It is conventional, helpful to the programmer, and helpful to any collaborators or others who happen upon a program later, to use this file extension for Python programs. Unfortunately, and confusingly, both Mac OS X and all recent Windows operating systems hide these extensions when showing the contents of a folder in a window. When you check via the command line, however, you'll see the full file name, with the extension. You'll also need to refer to the file this way on the command line. There are three approaches to dealing with this: you cannot care about the inconsistency, you can reconfigure Mac OS X or Windows to display file extensions by changing the appropriate preferences, or (the solution I prefer) you can replace your current operating system with GNU/Linux. That's not a joke; I would consider it, even if you don't want to make a radical change of this sort right now, while you are starting to study programming.

Save the file hi.py in the directory (or equivalently, folder) that you've created for your work as you go through this book. On my system, a GNU/Linux system where my username is nickm, the file ends up being /home/nickm/Desktop/explore/hi.py. This full, long identifier for the file is called the *pathname*. Although it will be different on Mac OS X, different GNU/Linux distributions, and Windows (which among other things uses backslashes instead of slashes) and will be different depending upon one's user name, all systems will have a similar sort of pathname for this file after it has been saved in your directory.

I use all-lowercase file names and directory names throughout this book for reasons of consistency and convenience. If everything is entirely in lowercase, we never have to worry about how a particular file name is capitalized when we are typing on the command line. Also, all-lowercase names are slightly easier to type because it's not necessary to hold down Shift. This is not the only way to name files, and it's even in conflict with conventions for naming files in certain programming languages. As you continue to work on larger projects, you should follow the standards of the language you are using and of the collaboration in which you are participating. Initially, though, it's most important to realize that capitalization matters and that conventions are established for good reasons. For instance, while I was writing this book, I created the following four files on my system in the same directory:

```
hi.py  hi.PY  HI.py  HI.PY
```

These are four distinct files; their names are obviously not much help in determining what they hold and which one is the version I'm interested in. Consistently using all-lowercase file names eliminates this one particular type of confusion.

Here is an operation of a few seconds that you should quickly undertake: using your system's graphical user interface, open your file viewer, open your home directory (on Windows, the *user folder* is the more common term), and navigate to your equivalent of the explore folder within that. You should see the file hi.py (or hi, if your system is set to conceal extensions from you) presented in this familiar view.

Now, with reference to the instructions in 2.4, Find the Command Line, open a terminal or command prompt so that we can do the same thing using this text-based interface. If you are running Windows, be sure to open Anaconda Prompt (not the default Command Prompt). The command-line interface to your computer can be used to navigate through your file system (as you just did in your GUI); to program in many different languages; to display, search through, and edit text files; to maintain and update your operating system; to check email; and to connect to multiuser environments such as MUDs and MOOs. As a starting point, we'll revisit how to use the command line to move through the file system and we'll use it to run Python programs.

There are all sorts of ways that graphical applications signal what they are so that you don't confuse your video editor with your Web browser and start trying to use one as the other. On the command line, the basic user interface is the *shell*, although some programs take over the whole textual window and a have a distinctive text border. The *standard* Python interpreter is such a program, but for our work we use the browser-based Jupyter Notebook.

Just as the top window of your file viewer shows a particular folder's contents, your shell is also context-dependent. Your commands apply based on your current location within the hierarchy of your computer's file system. It's possible to refer to any file at any point, but if you're interested in working with files in a particular directory, it's easiest to *change directories* so that you are "there" and can refer to the file you are interested in simply by typing hi.py instead of typing the full pathname. You probably will want to run modified versions of this file, or other files in the same directory, many times, so changing to the relevant directory once is efficient in terms of your typing. It's also less error-prone.

In what follows, I'll assume that you have a folder on your desktop called explore.

GNU/Linux or Mac OS X: To determine your current directory, type pwd (for *print working directory*) and press Enter/Return. (You need to press this key after any command entered in the terminal.) When you open a terminal, initially and in most cases, the current directory should be your home directory. That's /home/nickm for me; for you it will probably be /home/ followed by your username. At any point, cd ~ followed by Enter will change to your home directory. The ~ symbol is a shorthand for this directory. From there, it's easy to enter the Desktop directory (which contains what you see on your desktop) by just typing cd Desktop. Next, entering the explore directory is done in the same sort of way: cd explore. Now, pwd will reveal your current directory—hopefully the one you expect to be in (if not, redo these steps starting with cd ~). Typing ls (*list directory contents*) will give you a text-based view, within the terminal, of what you can also see by clicking on the explore folder on your desktop.

Here are the commands in order; you will have some sort of prompt, such as $ or perhaps something longer, at the beginning of each line, but what I've listed ahead is what you should type. I've added comments, beginning with #. You can type these in if you like, but it's not necessary; they have no effect on what the commands do. You do need to press Enter after every line, of course.

```
cd ~

pwd # Tells you which directory you are currently in, doesn't "move" you.

cd Desktop

ls # Lists all files, including directories, on the desktop.

cd explore

ls # Lists all files in "explore."
```

Great—don't close that terminal. Skip ahead past the next paragraph with instructions for Windows users.

Windows: When you run cmd from the Start Menu's Run option, a window will open with a prompt that indicates your current location, your user directory. It will contain a directory name similar to C:\Documents and Settings\Nick Montfort. I won't repeat this prompt (which will be different for you, anyway), but type the following to see the contents of the current directory (using the dir command), change to your desktop, see the contents of your desktop, change to the directory named "explore" within it, and check the contents of that as well:

```
dir
```

```
cd Desktop
```

```
dir
```

```
cd explore
```

```
dir
```

The main difference here is that `dir` is used in Windows to see what's in the current directory, while `ls` is used in GNU/Linux and Mac OS X.

On all platforms: One special sort of directory indicator is ~ and another one is ..— that is, two periods in a row. The ~ indicates your home directory or user folder. The two dots mean *parent directory*, the directory that is hierarchically above (and that contains) the current one. Try `cd ..` followed by `ls` (or on Windows, followed by `dir`) to see that this first command moves you up, back to the desktop.

Now that the current directory is the one we named "`explore`," we can run the program that we're interested in without typing some elaborate pathname such as /home/nickm/Desktop/explore/hi.py or perhaps something even longer. It's a simple matter of entering:

```
python hi.py↵
```

This is the same `python` command that can be used by itself on the command line to start the standard Python interpreter, although we are using the more powerful Jupyter Notebook. Here, we add something to this command—an *argument*. This is an extra piece of data for the `python` command to work with. Although the shell and Python are not the same environment, this command-line argument is really quite similar to the argument that "Double, Double" accepts within the interpreter, within Python. In the case of "Double, Double," the argument is the thing to be doubled, a sequence of some sort, such as the list [`-42, 0, 42`]. The argument to `python`, on the other hand, is a file name, the name of a Python program. Instead of starting an interactive session, Python runs this program, presenting the output in the terminal window. Then we are back in the shell, so commands such as `cd ~`, `ls`, and `pwd` can once again be used at this point.

Once you've typed some commands into the terminal, you may find that you wish to issue a command again. For instance, you may wish to run the hi.py program again. To do so, just press Up Arrow (↑) followed by Enter. Up Arrow brings the last command that you used back onto the current line. You can use Backspace to edit it if you want, but for now, Up Arrow and Enter will serve perfectly well to run the "Hello World" program again. You can also move up and down through your history of commands with the Up Arrow (↑) and Down Arrow (↓) keys. This means of navigating through previous commands works in the GNU/Linux and Mac OS X shell, the Windows command line, and even in the terminal-based Python interpreter, should you choose to use that at some point.

Using your text editor, edit hi.py and replace "Hello World" with some other string—keeping the quotations marks and everything else the same. Save the file (just overwrite the previous version this time) and go back to the terminal. With Up Arrow and Enter, you can run hi.py again. (If you've typed some command after `python hi.py`, you will have to press Up Arrow more than once until you find the right line—or, of course, you can retype the command.) This time, the file you run will be the new one, and you will be presented with the new string.

At this point, without having discussed the specifics of programming much at all, we already have two ways to run small, simple programs—in both cases, programs in Python. We can open the interpreter and type programs in directly. (Not every programming language allows this, but several others, such as Ruby and Lisp, do.) We can also write a program in a text file, save it, and run it from the command line. (This is a widespread way of writing programs, particularly simple ones.) In JavaScript work, we edited text files in a very similar way, but to run the programs we opened our files, with scripts embedded in HTML, in a Web browser. In working with Processing, as mentioned before, we will use that language's elegant IDE rather than a text editor.

It takes some effort to gain familiarity with the command line. The Terminal or Command Prompt is an entirely different way to access your computer—more powerful in some ways than the GUI and not based on all of the same principles or metaphors. Still, when you use it, you're accessing the same file system, organized the same way. And you can start with a few commands (`ls`, `cd`, `python`) and your existing understanding of how computers and their file systems work as you use the command line. From there, you can learn enough to use the shell in ways that help you. This much experience may not enthrall you with the power and flexibility of the command line, but I hope it will at least make it a bit less mysterious.

While "Hello World" (with or without a comma or exclamation point, capitalized variously) is the most famous first text to have the computer display, alternate traditions have arisen. At the Stanford Artificial Intelligence Lab (SAIL), a different text was sometimes substituted: "Hello Sailor." While helping establish a distinct body of Stanford computing lore and customs, this was also a loaded phrase, evoking a sexual proposition that was often thought to be used by prostitutes soliciting nautical customers. It may also suggest a sexual solicitation by someone, male or female, who is not a prostitute. If "Hello World" suggests the wonder of Miranda, finally in contact with the amazing wide and inhabited environment, "Hello Sailor" suggests a very different, jaded perspective—and perhaps a campy one. The phrase is well-known enough to have been used as a book title more than once, in the 1997 book *"Hello, Sailor": Changing Perspectives of Japanese Prostitution: A Historical Analysis* and the 2003 book *Hello Sailor! The Hidden History of Gay Life at Sea.*

There are references to "Hello Sailor" and this sense of the phrase in creative computing, too—particularly in the Infocom games in the *Zork* series, based on a minicomputer game written from 1977 to 1979. The phrase occurs throughout the series, perhaps most memorably in the black book found in *Zork I*, where the text begins by admonishing the reader:

Oh ye who go about saying unto each other: "Hello sailor":

Dost thou know the magnitude of thy sin before the gods?

The rest of the text describes the punishments that those who say this phrase will face, which correspond to incidents in Odysseus's journey. *Zork* shows a clear awareness of this phrase's sexual sense; it also pokes fun at those from SAIL, the Stanford lab where Don Woods, using original code by Will Crowther, completed the canonical version of *Adventure* (1977), a game also known as *Colossal Cave*. This game was the direct ancestor of and inspiration for *Zork*. Since it is necessary to say "hello, sailor" later in the *Zork* series, in *Zork III*, in order to solve a puzzle, it would be difficult to consider that this religious condemnation is meant seriously.

It's easy to imagine other things that "Hello World" might say. A relatively innocuous one, suggesting the movie *Tron*, is "Greetings, Program." A phrase can easily be imagined as part of a program of this sort. But it's also easy to actually try out what it's like for the computer to say something else, and even this simple of an exercise can be informative in several ways.

Trying to have the computer display "Buenos días," for instance, will reveal whether the programming environment you are using supports accented characters from the Latin alphabet. It seems like it should, at this point in the twenty-first century, but it might not. Trying to have the computer display the greeting "привéт" will reveal whether one's environment supports Cyrillic. (If you want to do this sort of test this without changing your keyboard layout, you could paste in a Russian word, in Cyrillic, from the Web.) Other characters, including ones from non-Western writing systems, can be tested in this way too. With Unicode support, a wide number of glyphs may be supported, but it may not be equally easy to type all of them. All characters are equal, no doubt, but you will probably find that some are more equal than others in your system.

Actually trying out different texts, rather than musing about what it would look like for a computer program to produce those texts, can be a useful exercise in other ways. It's easy to imagine provocative, offensive, and subversive texts to use, but actually putting these into a program, or trying to, can help the programmer discern which are trivial or juvenile and which are truly charged. Can your computer provoke and offend just by printing a string? Or does it lack the social standing to do so? Who will end up bearing the blame (you, as the programmer, or the computer, as the one speaking) if your program does say something offensive?

[Exercise 7-1] Rewrite the Greeting

Replace the string "Hello World" with some other text and share your program with others. This can be done by everyone in a computer-supplied classroom within a few minutes; those studying alone can try out their alternative starter program on themselves and on friends familiar with "Hello World." In doing this exercise, you also show your ability to run text-file Python programs from the command line and your ability to modify an existing Python program.

[7.3] Temperature Conversion

Let's return to trusty Jupyter Notebook for this next starter program. You can open up a new notebook, as it shouldn't be particularly important to refer to or rerun previous code. By the way, even if you are writing a Python program in a text editor, it's perfectly fine—probably even a good idea—to have a Python interpreter of some sort open so you can try out short code snippets and confirm your understanding of how Python works. Running both a text editor and Jupyter Notebook is extremely unlikely to tax your computing resources, but it is likely to make programming easier.

Converting between Fahrenheit and Celsius is often the first mathematical function implemented by new programmers. A conversion program demonstrates how computation is general—how it can work on arbitrary values. But this is also a very practical program, for an important reason: measurement systems are culturally situated and, when people encounter a temperature display or a fixed value from a particular different culture, it is necessary to convert from one scale to another.

The United States and a very small number of other places still use the Fahrenheit scale. Let's consider what some code to convert a "rest of world" Celsius measurement into Fahrenheit *should* do—keeping in mind that this is our specification; it will not work right away and should not be typed in right away:

```
to_f(27)
```

That sort of function would be fine, although we haven't written the function yet. For this particular value, the corresponding Fahrenheit temperature is 80.6 degrees, and the number 80.6 would be an appropriate result. We would provide whatever Celsius temperature as a numeric argument to this function, and the function would return the converted value.

The Celsius measurement system, devised by astronomer Anders Celsius, seems to be particularly well-calibrated in a culturally neutral way. The scale was set (until the middle of the twentieth century) so that 0 is the freezing point of water and 100 is the boiling point. Water freezes and boils at different altitudes and pressures, so sea level and mean barometric pressure was specified. This scale was first formulated in 1742.

The Fahrenheit scale, which is slightly earlier, is also calibrated with regard to the natural world, but somewhat differently. Daniel Gabriel Fahrenheit, the inventor of the glass-contained mercury thermometer, devised the temperature scale that bears his name. In 1724, he determined that 0 would be set to the temperature which was thought, at the time, to be the coldest possible—the freezing point of brine, or specifically of a mixture of equal parts ice, water, and ammonium chloride, which stabilizes at a particular temperature. The 32-degree point was the freezing point of water (measured by the same process, just without the ammonium chloride), and the 96-degree point was set at the temperature of a healthy human body, or "blood-heat." The scale was later recalibrated, but this seems to be how it was set up initially. Although we might consider a scale running from 0 to 100 as more obviously useful today, the 64 points between 32 degrees Fahrenheit and 96 degrees Fahrenheit were carefully considered in the construction of the scale. Fahrenheit could simply determine the midpoint of that interval, physically mark on glass, and then divide each segment in half again. After six such divisions, he would have all the markings from 32 to 96 made.

The Fahrenheit scale is offset 32 degrees because the freezing point of water is 0°C and 32°F. Between freezing and boiling, there are 180 degrees on the Fahrenheit scale (from 32 to 212) and, of course, there are 100 on the Celsius scale. So, to convert, we simply multiply the Celsius temperature by 180/100 and add 32. We can simplify this: 180/100 is 9/5. Thus, we need to multiply whatever value is given by 9/5 and then add

32 to that. Fill in this partially completed template, a sort of program-in-progress, so that you have a function that works properly:

```
def to_f(c):

    return _____↵
```

Try this out on some values, such as the freezing point and boiling point of water and some "room temperature" value.

Prove to yourself that this to_f() function allows us to convert any Celsius temperature to Fahrenheit. The only complaint that could be raised against this function is that it will convert values below absolute zero. We'll deal with that complaint soon.

How about converting Fahrenheit values to Celsius? Let's consider what our current function does and get a bit algebraic on it. It computes $((9/5) \times C) + 32$ and returns that value; that is, $F = ((9/5) \times C) + 32$. To write to_c(), we just want to turn this around, treating it as if it were an equation and solving for C. For a first step, subtract 32 from both sides: $F - 32 = (9/5) \times C$. Then, multiply both sides by $(5/9)$, as in $(5/9) \times (F - 32) = C$. This tells us how to write the function we're interested in:

```
def to_c(f):

    return (5/9) * (f-32)↵
```

This is a seemingly trivial bit of computation. However, it does a conversion that proves difficult for many people to do quickly. This conversion can be quite useful when one is in a different country. The conversions we've implemented are general; they work on any temperature. And they're packaged in functions so that instead of typing everything again and again, and possibly making errors, we can simply type things like to_f(25) and to_c(92).

Now, it seems that a clever and effective way to test our two programs would be to see if to_f(to_c(25)) was (at least approximately) 25 and if to_c(to_f(90)) is 90 and so on. We'd like to call one function on the other and vice versa. Try that out and see how it works.

By the way, if we had tried to print our result instead of returning it, we couldn't use this method of testing the programs. Go ahead and replace the return ___ with print(___) in both functions and press Shift-Enter to confirm the change you've made. Now try, for instance:

```
to_f(to_c(25))↵
```

This is why we can't have nice things if we write our functions to print() (i.e., to visually display) values instead of using return so that other code can make use of the result. Go ahead and change `print` back to `return` in both functions, following best programming practice.

While Celsius certainly seems to be the more reasonable and well-grounded temperature scale, the Fahrenheit scale has some advantages for everyday use. If a person is wondering what temperature it is outside, it's usually enough, in Fahrenheit, to know that the temperature is in the twenties, thirties, forties, fifties, sixties, seventies, eighties, or nineties. A single digit (the first one) captures enough to inform someone about what to wear. In Celsius, that range extends from about -7 to 37, and it's not very helpful to know only the first digit of the temperature. So, awkwardly calibrated as the Fahrenheit scale is, it arguably is better suited to at least some of our common experiences of temperature, at least in this one way. One could make the argument that it's on a more human scale, good for distinguishing between temperatures that we usually encounter.

The specifics of these temperature scales are somewhat interesting, but perhaps most interesting is that they, like so many measurement systems, do not exist as mathematical abstractions. They are determined by convention, law, and national norms. As soon as we encounter the power of the computer as a pure calculating machine and begin to imagine its rarefied existence apart from matters of mere humanity—we are shown an example of how computation can be used for mediating between the cultural norms of different measurement systems. While we might want to imagine detached, abstract, perfect and pure logic and mathematics, what we have instead is Daniel Gabriel Fahrenheit slapping a thermometer up against a healthy man's armpit.

[7.4]   Lowering Temperature and Raising Errors

You have seen that syntax errors can arise as you're writing programs. This Celsius-to-Fahrenheit conversion task provides an opportunity to show how you can write your code to raise errors when your program has detected that some mistake must have been made. In this situation, we will have our temperature conversion program make use of scientific results. Absolute zero, 0 Kelvin, is –273.15 degrees Celsius, or –459.67 degrees Fahrenheit. Any lower temperature value is nonsensical: it cannot correspond to anything in the universe. If someone asks for a larger negative value than –273.15 to be converted from Celsius to Fahrenheit, or a larger negative value than –459.67 to be converted from Fahrenheit to Celsius, it must be a mistake. Whatever number they are using as an argument could represent something else—how much money someone lost gambling, for instance—but it can't be a temperature. So, we'll have our program explicitly say that the value provided is an erroneous one.

Let's start by modifying the short Fahrenheit-to-Celsius program in Jupyter Notebook. It looks like this to begin with:

```
def to_f(c):

    return ((9/5) * c) + 32
```

We'll simply add one conditional statement:

```
def to_f(c):

    if c < -273.15:

        raise ValueError("Temperature value is below absolute zero.")

    return ((9/5) * c) + 32↵
```

If the condition is true, which happens whenever a bogus value is supplied, an error will be *raised* (that's the term for it) and execution of the program will stop right at that point. The program will not proceed to the next line to return a value.

The specific type of error will not be a SyntaxError because there is no problem with syntax here. Instead, it will be a ValueError, which you can use for any sort of inappropriate value—for instance, if somewhere in your program you found that a person was indicated as working negative three part-time jobs. Because a string has been given as an argument to ValueError, a helpful message will be produced if this error is raised.

Go ahead and try the new to_f() function with a few values, some valid and some invalid.

Once you see how this works, fix your to_c() function that converts Fahrenheit temperatures to Celsius. Add code so that this function raises the same error under the correct circumstances. Test the function to make sure it works as it should, producing an error when an error is appropriate.

[7.5]   Converting a Number to Its Sign

In the previous chapter, in 6.8, The Conditional, if, the function yesno() was introduced. This short function returns True if its argument is 'yes', False if its argument is 'no', and nothing otherwise. This function may not seem very similar to one that does Fahrenheit-to-Celsius conversion, but it is also doing a type of conversion. At least, it's possible to think of this function as doing conversion. It maps one string to True, another string to False, and all other strings to no return value at all.

Now we'll write a slightly longer type of categorical conversion function. This one will tell us the sign of a number: '+', '-', or '' for no sign at all. This one should be a *total* function, which means that there is a return value for every possible number given as an argument. One of those return values will be the null string, the string with zero characters, but that's still a return value. Type this into Jupyter Notebook:

```python
def sign(num):
    answer = '?'
    if num > 0:
        answer = '+'
    if num < 0:
        answer = '-'
    if num == 0:
        answer = ''
    return answer
```

This program first assigns '?' to the variable answer. Then it goes through and checks three conditions. If the argument is greater than 0, '+' replaces '?' as the value of answer. Next, if the argument is less than 0, '-' is assigned instead of '?' as the value of answer. Next, if the argument is equal to zero, the empty string is assigned to replace '?' as the value of answer because zero has no sign. And finally, the function returns the value of answer, whatever it is.

To try this function out:

```python
sign(4)
```

```python
sign(-20)
```

```python
sign(0)
```

```python
sign(1)
```

This program is written with a single return statement at the end, which is generally considered good practice. It would be possible to do away with the variable answer and simply return '+' or '-' or the null string at the points in this code when those values are now assigned to answer in the code. This results in a program that is two lines shorter. That result is not an incorrect program. Go ahead, try editing this version down in this way and then seeing how it works. Because every case is covered (the argument is either going to be greater than zero, less than zero, or zero), there's no problem with this shorter version.

While you're at it, also try out this version of the code, which seems to me to be an improvement:

```
def sign(num):
    answer = '?'
    if num > 0:
        answer = '+'
    elif num < 0:
        answer = '-'
    else:
        answer = ''
    return answer⏎
```

Notice that this version uses elif and another different keyword, else. The last branch of a conditional can be indicated with else, and if it is, it will be taken when none of the earlier conditions hold. It means "otherwise . . .". The elif combines an else and an if, indicating "otherwise, check to see if . . .". In this case, the conditional tests to see if the number is positive; only if it isn't does it proceed to test whether it is negative, and then, if it isn't negative either, without doing any additional tests ("otherwise . . ."), proceeds to the case that applies when the number is zero.

This is a situation (one of many) where you can use good programming practices or not, given the immediate problem at hand. Still, good practices can pay off in other situations. To explain this, I'll introduce a technique that will be discussed more fully at the end of 8.3, Selecting a Slice: the way to get the last letter of a string. This is done simply by placing [-1] after a string variable or after a value such as 'hello'. Try it in Jupyter Notebook:

```
name = 'Paula'
name[-1]⏎
'a'
word = 'hello'
word[-1]⏎
'o'
'world'[-1]⏎
'd'
```

Now consider another function, `gender()`, that takes a string as its argument. This function determines what gender a particular name signifies, based on the special naming system used in the bifurcated, imaginary land of Binaria. In this country, which uses the Latin alphabet, all boys are given names ending in *o* and all girls are given names ending in *a*. According to the strict rules of Binaria, only two genders are possible. This is a land of twofold determinations. Here's the function for you to type in:

```
def gender(name):
    if name[-1] == 'a':
        return 'female'
    if name[-1] == 'o':
        return 'male'↵
```

With that function defined, we can type:

```
gender('Apollo')↵
```

That invokes the function, and Jupyter Notebook tells us what its return value is. To update this function with a new version after we've changed something, we can simply copy the new version from the text file and paste it in.

If you try this function out on "Francisco," "Clara," "Gonzalo," "Ola," "Roberto," and "Vanessa," you may be perfectly satisfied with it—given the distinction that it is supposed to make. But perhaps Binaria has some immigration from places such as the United States, or requires visas for visitors, or something like that. Because of that, a list of names in Binaria may at times contain some names that do not end in *a* or *o* as all native names do. For instance, let's say that I applied for a Binarian visa and did not enter my first name as something like "Nicko" or "Nicka." If you try `gender('Nick')`, you'll see that the function, with this argument, returns nothing at all. This could cause problems for me if I need to use the bathroom.

I'm trying here to introduce an issue that is serious, although with a made-up country and a situation that may seem silly. Computers have been used to formalize and maintain traditional categories, including binary gender distinctions, ever since computers were invented. While this might sound like a joke when it's part of a contrived example like this, people who do not identify as one of the two standard genders concretely suffer from these sorts of binary classifications all the time. This isn't a new issue; there is a long global history of people who don't wish to be thought of as male or female. This matter has come to broader public attention in the United States and other countries recently, however, and more people are open about their discomfort with or opposition to the binary system of gender. Understanding how computer programs categorize, and how to build new and different systems, can have positive social potential when a classification issue like this one is involved.

We will try to make a tiny improvement, at least. Here's another version of the function, similar in structure to `sign()`. This one still follows the rigid rules of Binaria but is potentially improved in some ways:

```
def gender(name):
    indicated = '?'
    if name[-1]  == 'a':
        indicated = 'female'
    if name[-1]  == 'o':
        indicated = 'male'
    return indicated⏎
```

At the cost of just a few additional characters, this function is a bit more honest about what it is doing. If any string is passed in that doesn't end in an *o* or an *a*, the function now returns a question mark to indicate that it cannot determine the gender of the name using its rule system.

There's also a new variable in this longer function, one named `indicated`. This is hardly a clear and well-commented function—it doesn't actually have any comments at all—but thanks to this variable and its name, there is at least more of a clue that the function is figuring out the gender *indicated* by the name, according to Binaria's traditions. This may not be the same as the gender actually assigned to an individual by society or the gender an individual might choose in terms of self-presentation.

[Exercise 7-2]   A Conversion Experience

Create another simple function that converts between units of measurement of any sort. They can be units of practical importance if you like. They might be culinary units such as US tablespoons and Australian tablespoons (which, indeed, are different). They might be computational units such as the gigabyte (when defined as one billion bytes) and the other gigabyte, aka the gibibyte (1024^3 bytes, which is 1,073,741,824 bytes). Or they might be more fanciful units, such as human years and dog years. Instead of using `print()` to display the result when your function is finished, `return` it. This should be a very simple and short program that works much as `to_c()` does. Work in Jupyter Notebook to develop and test your function. Then, as a final step, place your function in a text file with some test cases at the bottom; those test cases should involve calling your function with different arguments and using print() to display the result.

[Exercise 7-3]   Categorical, Imperative

Create another simple function, along the lines of sign() and gender(), that classifies its argument into one of a small number of categories, such as three. You are encouraged to select categories creatively and are required to keep the program very short (ten lines or less). Try to determine some interesting categories that can be easily discerned by a computer, based on what you know about computation so far. For instance, while it may be interesting to distinguish prime numbers and composite numbers, we have not discussed (and are not going to discuss, in this book) how to determine whether or not a number is prime. So it would probably be a better idea to determine a simpler type of categorization, perhaps based on whether a value is, for instance, equal to some meaningful value or not or whether it is greater than such a value. You could examine words using len(), to see if they are short, of medium length, or long, or you could determine whether numbers are even or odd. If you find yourself writing an elaborate test, choose something simpler to categorize!

[7.6]   The Factorial

Computing the factorial is a reasonably common thing to do when starting to learn various programming languages.

The factorial of a number n, written $n!$, describes how many ways n item can be arranged in a sequence. In other words, $n!$ is the number of permutations of n. For instance, 1! is the number of ways that a single item can be arranged—only one way, of course, so 1! is 1. Next, 2! is how many ways two items can be arranged: they can be set up as 12 or 21, two different ways. For 3!, we can pick any of the three items for the first spot, either of the two remaining items for the second spot, and then we don't really have a choice for the third spot, which will have to hold the one remaining item. So 3! is $3 \times 2 \times 1 = 6$.

In fact, $n!$ can always be computed as n multiplied by n-1 multiplied by n-2 all the way down to one.

Knowing this, let's write a function, using iteration, that will compute the factorial of any number n that is greater than or equal to 1. First, let's see how we can get a sequence that begins with 1 and continues 2, 3, 4, up to whatever number we like.

Try typing the following in Jupyter Notebook:

```
list(range(5))↵
```

The outer function, list(), is a built-in function that makes a range into a list; it converts its argument to the list type. Without getting into unnecessary detail, in Python 3, we need to use list() to see all the elements that range(5) produces, all at once. You'll see that this returns [0, 1, 2, 3, 4], a list that is almost exactly what we are looking for—but not quite. This list starts at 0 and goes to 4. This is quite handy if we're trying to go through another list by index, because, as we'll see soon, a list with five elements begins with index 0 and ends with index 4. However, it's not what is needed to compute the factorial of 5. In that case, we need to compute $1 \times 2 \times 3 \times 4 \times 5$, not $0 \times 1 \times 2 \times 3 \times 4$.

This range can be specified with two numbers instead of one, a starting point and the number before which the sequence will stop. Notice that the following use of range does the same thing as the previous one; it's just a way of rephrasing it:

```
list(range(0, 5))↵
```

Now, let's have the range begin at 1 and go up to (but not include) 6:

```
list(range(1, 6))↵
```

That's the list we're looking for. This is a great time to experiment with range(). Think of any consecutive sequence of integers, such as [15, 16, 17]. (A short sequence is fine.) What first number and second number for range() will produce the sequence that you have in mind? Try out a few different uses of range() so you get a feel for how the first argument is the starting point and the second argument is the first value that *doesn't* appear.

Now, let's consider how a factorial function will work. It needs to multiply together every number in the range that begins with 1 and ends with *n*. So, we start by assigning the variable answer the value 1 and we keep accumulating the answer in this variable, multiplying by each number in turn and storing the result there. Finally, when we've done all the necessary multiplication, we return the answer.

Let's first do this process ourselves, without constructing a function. We'll compute the factorial of 5. So that we know what to expect, consider that $5 \times 4 \times 3 \times 2 \times 1$ is $(5 \times 4) \times (3 \times 2)$, which is 20×6, which is 120. If we think about it deliberately in this sort of way, we don't need a computer to find the result in this case, but we have to check that our method works using values we can compute by ourselves. Try this in the interpreter:

```
answer = 1

for num in range(1, 6):

    answer = answer * num

    print(answer) # Display the current value

answer↵
```

On the first line, we're just assigning the variable answer the value 1. The next line loops through our sequence from above: [1, 2, 3, 4, 5]. It multiplies answer by each value in turn. So on the first iteration, answer is 1. On the second, it's 2. On the third, it's 6. On the fourth, it's 24. To be able to see this visually, I have included another line of code that is indented and thus within the loop, the print() function. This simply provides an informative display, showing the current value of answer during each pass through the loop. Be aware that print() is not returning a value; it's just giving us a sort of status update so we can watch the program's progress as it proceeds. After the fifth and last iteration, the answer is five times the previous amount, 24, which ends up being 120, the desired result. That value is actually returned, rather than just being displayed.

As I see it, this is another type of "accumulating," some sorts of which we have seen before. Instead of adding at each step, the code multiplies at each step, but is similarly working to build up the final answer.

All that remains is to encapsulate this code in a function and generalize it so that instead of just working for the value 5, it works for any positive integer. Now that the code is working, I will remove the invocation of print() so as not to clutter things up. Here is most of how to encapsulate this process in a function:

```
def factorial(n):

    answer = 1

    for num in _____:

        answer = answer * num

    return answer↵
```

The blank is where you will need to specify the range so that the iteration begins with 1 and goes up to (and includes) the value of *n*. Once you determine that range and place it there, you should have a working factorial function. Go ahead and figure out what goes in the blank and place it there.

A fun fact: your function actually works not only for 1 and all larger integers, but also for 0. The idea of 0! may seem a bit perplexing at first, but it's defined in the same way that the factorial of other values is defined: it's the number of different ways that zero elements can be arranged in a sequence. How many ways is that? It's the number of ways you can place zero books on a shelf: just one way. So, given that definition, consider what happens when one invokes `factorial(0)`. If the `factorial` function is given 0 as an argument, the answer should initially be assigned the value 1, the function should go through the for loop zero times, and the answer that results at the end should be simply the original 1.

We now have a good factorial function that will work with any whole number. As a next step, we'll write a different function to do the same thing.

Some people who teach programming hold that it's a bad idea, very early on, to show multiple ways of solving the same problem. I find it important for both artists and humanists to understand that different code-level choices can be made to accomplish the same thing. The recently established fields of critical code studies, software studies, and platform studies rely on understanding how there are different implementations for what is operationally the same program. These code-level differences have many sorts of cultural importance. And as someone who seeks to think and explore using code, it's important to see how one's explorations involve choices and expressivity, how they aren't simply rote mappings from a problem to a single solution.

So let's consider a different approach to the factorial function. I sketched out how the factorial of n can be computed by multiplying all the values between 1 and n. But there is another way of thinking of this function. We could also say that the factorial of 0 is 1 (as noted in the fun fact earlier), and the factorial of every larger number n is $n \times (n - 1)!$. That is, we'll say that 5! is 5 × 4!, and 4! is 4 × 3!, and so on. This is really just another consistent way of phrasing the definition we've already seen.

To keep things straight, we'll call our new function `fact()`.

```
def fact(n):

    if n == 0:

        return 1

    else:

        return n * fact(n-1)↵
```

This function is quite different. It does not iterate; there is no for loop. Instead, there's a single conditional statement. In the first case, if n is 0, the function returns 1, the correct answer.

In the second case, the function returns n multiplied by whatever the value of $(n - 1)!$ is. That value is computed using *recursion*: the function calls *itself* with a different argument, one that is always getting smaller.

If it seems odd for a function to call itself, why didn't it seem odd that we defined *n*!, just a moment ago, as *n* × (*n* - 1)!—that is, that our earlier definition of factorial was in terms of the factorial function itself? In our mathematical definition and in our function, we are always checking to see what the factorial of a smaller number is. If we start with a positive number, we'll eventually reach zero, which has a defined value. And at each step down, we're multiplying by the current value of *n*, so we reconstruct the expression *n* × (*n* - 1) × (*n* - 2), which we got earlier using iteration.

You should note that it's very important to ensure that this process eventually stops! If we didn't have the first part of the conditional, checking to see if we've reached 0, we would end up trying to build an expression with an infinite amount of terms. Trying to compute 3! would involve trying to compute 3 × 2 × 1 × 0 × –1 × –2 . . . and so on, forever. You can see this catastrophe happen (don't worry, it won't destroy your computer) by typing in the following and testing it out in Jupyter Notebook. Type in the following and test your new function on a number such as 3 or 5:

```
def factfail(n):

    return n * factfail(n-1)↵
```

Once you understand the basics about what recursion is, the error message is fairly informative, isn't it? Here you'll see a twist on the semantic error, the non-intentional program. The function is entirely valid, but produces a runtime error as it exhausts your computer's resources. It can't return any answer at all, even the wrong one.

The recursive way of computing factorials will probably present itself as the "obvious" one to those who prefer to program in a functional style, using languages such as Lisp and Haskell. The iterative way is perhaps better suited to those who prefer imperative programming, which is often done in C, Python, and Java. There are some low-level advantages and disadvantages to each, but for thinking, sketching, and exploring, the important thing to understand is that both ways of thinking about problems do indeed exist, and both are useful.

[Exercise 7-4]   Negative Factorial Fix

You may have noticed that if you give a negative argument to `factorial()`, the result will be 1, which is not the right answer. The factorial of a negative number, like the result of dividing by zero, is undefined. Elaborate the conditional statement so that if a number less than 0 is given as an argument, `factorial()` will raise a `ValueError` as described in 7.4, Lowering Temperature and Raising Errors.

Having done this, turn to `fact()`. This function will actually crash if given a negative argument, for reasons just explained in discussing `factfail()`: once this function ends up "below" the base case, the limit of 0, it will keep subtracting one from that argument and calling itself until memory has been exhausted. Add a conditional statement so that if a number less than 0 is given as an argument, fact() will also raise ValueError.

[Exercise 7-5]   Factorial Mash-Up

As written, fact() calls itself and factorial() does not call any other function. Could fact() be written to call factorial() instead of itself? Would the modified version of fact() correctly compute factorials? Why or why not, exactly? Write a statement about why you expect this modified function to work or to break. Then, change fact() so that it invokes factorial() rather than itself and see what happens. Were you right? How much testing did you need to do to determine whether or not the new version of fact() works? This exercise is not meant to show you that it's a good idea to write two different implementations of the same operation and then hook them up to each other. That's bad programming practice! This, however, is an exercise in understanding abstraction and the interface to functions.

[7.7]   "Double, Double" Again

The alternative "starter program" presented in this book is double(), which was developed to serve as an example of several computational fundamentals: functions, iteration, and polymorphism. It is meant to pack a lot of the abstraction that can be accomplished by computer programming into a few lines of code.

Whereas "Hello World" is a stand-alone program that can run and produce output by itself, "Double, Double" is a function, similar to Celsius-to-Fahrenheit conversion and the factorial function. It maps input values to output values. It works on sequences of data but does not perform a culturally meaningful translation, as the temperature conversion program does. The factorial function probably provides a more "natural" example of recursion versus iteration, although double() can also be implemented recursively. Let's see how.

To display twice the value of every element of a list, we mentioned that we could simply have the program go through each element (iteratively, using a for loop) and print twice the value of each one. Of course, that works.

But we can also think about a list differently. Let's consider the length of the list: this length is either 0 or some number greater than 0. If the length is 0, then we have a list with no contents, the empty list. We have zero elements to double. The result should be the same thing we started with, the empty list.

If the length is greater than 0, then the list must have a first element followed by what we'll call the rest of the list. (Even if the list is only one element long, it still has a first element and a rest of the list. In this case, the "rest" is the empty list.) Now, the doubled list that we seek (a list with twice the value of each element) can be determined by adding together twice the first element and the double of the rest of the list.

The code for this is:

```
def doubler(sequence):

    if len(sequence) == 0:

        return []

    else:

        return [2 * sequence[0]] + doubler(sequence[1:])
```

Is this perhaps less clear than the previous version, double()? Because this is the first time we've used [0] to split off the first element of a list, and because [1:] may seem like a rather cryptic way to get the rest of the list, this version is probably not very clear yet. Whatever you think about this matter, though, both of these functions are reasonable expressions of two different ways of thinking about, and solving, the problem of doubling every element of a list. They both work, and they show one iterative and one recursive approach to a problem.

[Free Project 7-1]   Modify "Stochastic Texts"

Do this project at least three (3) times, developing different modifications. When I invited you to modify JavaScript programs, one of them was a classic, Theo Lutz's "Stochastic Texts" from 1959. I have made this program available, in reimplemented form, not only in JavaScript but also in Python. Find it here and download it:

nickm.com/memslam/stochastic_texts.py

Open the file in a text editor. Then modify it, as you did when you modified that JavaScript program, to make a text generator of your own that is based on this one. There are two goals here: First, you should get a bit more experience working with Python programs as plain old text files, using a text editor and running them from the command line. Second, you should try to make modifications to this program that are different in significant ways from any you may have already done in JavaScript. Try shifting to a different tone, or perhaps making the generated text more coherent or more uncanny.

[Free Project 7-2]   Modify and Improve Starter Programs

Do this project at least three (3) times, using two (2) different starter programs. Use one of this chapter's starter programs as a basis for your work and develop a modified program that you believe improves on the version presented here. As with the original program, the program you create should be simple and should do one simple thing. You can change the type of conversion done by the conversion program, however, or have the factorial calculator do something similar, but not exactly the same, when given a whole number argument.

[Free Project 7-3]   Write a Starter Program

Do this project at least two (2) times with different audiences in mind. Create an alternative starter program of similar simplicity to the four discussed in this chapter, this time starting from scratch and with your own idea. The program you create should do only one simple thing. In a short paragraph, you should justify this program as one that is good at being an initial introduction to programming and computation.

For instance, there was a 1994 *Saturday Night Live* skit, "Total Bastard Airlines," that presented the scene of passengers exiting an airplane. As the passengers left, members of the flight crew repeatedly said, "Buh-bye!" The actors playing passengers ran around the set, getting on the plane again and exiting the plane again, so that the flight crew was standing there and saying, apparently without end, "Buh-bye."

A suitable project could be a program that says—actually, displays on the screen, using print()—"Buh-bye" some very large number of times, using iteration. For those who want a "stretch goal," getting a program to say "Buh-bye" an *unbounded* number of times can be done but would require some additional exploration of concepts not yet introduced. The eternal "Buh-bye" program, with a justification for why it is a valuable introduction to computing, would be a fine free project that involves taking a step beyond what is outlined in this chapter. If you do it, you still need to justify it as a good starter program!

Whether you pursue this sort of project or another, remember that the goal is not to write an *elaborate* program, but to write a *very simple* new program that can be justified as a good introduction to programming and that does something computationally different from the programs introduced so far.

[Exercise 7-6]   Critique My Starter Programs

I provided three at least somewhat nonstandard starter programs, or ways of learning about computing: double() initially, in 5.2, Type In the Function, and then tax() and mean() in chapter 6. These are functions that don't do anything unless they are called, rather than entire stand-alone programs—but they serve to introduce computing, all the same.

Offer a critique of these. A critique doesn't require you to be negative, although you can be negative. To critique something, you just need to be discerning and conduct an inquiry. What does the program highlight about computing? What does it hide or obscure? Does it welcome people from different backgrounds to computing? Does it appeal more to those interested in some media, or particular studies and practices, and less to those interested in others? I was, of course, aware of some of the cultural implications of the standard starter programs, but how do the three I offered reflect cultural perspectives on computing? Having just developed your own starter program, you should have some interesting thoughts about this. Feel free to deliver your critique publicly, on your blog or favorite social media platform, and to invite others to the conversation.

Of course, the previous free project should have already given you the opportunity to deliver one type of critique, in code.

[7.8]   Essential Concepts

[Concept 7-1]   Computing Is Cultural
It would be so easy if we could just completely forget about the social, political, and cultural world when we think about computing and programming. However, the choice of what to program and the way in which programs work have significant cultural dimensions, just as architecture, design, art, and writing do. Be aware of the implications of this for one's work as an exploratory programmer—choosing what to explore, how to explore it, and whether or not you wish to allow others to use your code and build on your work.

[Concept 7-2]   You Too Can Raise Errors
Python produces several sorts of informative errors on its own. You can have code that you write produce its own errors—for instance, to indicate when a nonsensical value has been provided as an argument.

[Concept 7-3]   The Conditional Can Categorize
Having seen that the if statement allows for something to happen only if a particular condition holds, you should understand at this point how to sort data into any number of categories and how to specify a default value if the data doesn't fit into any of the categories.

[Concept 7-4]   Iteration and Recursion Can Both Work

Computing the factorial has given us the opportunity to see that some problems can be solved either iteratively or recursively. There will be other examples of this provided later, but you should grasp some of the basics of recursion at this point. Understand, for instance, that recursion involves formulating a solution as some partial answer combined with another application of the process. And understand that functions can in general call any functions they like—other functions, for instance, but also themselves.

[8]   Text I: Strings and Their Slices

[8.1]

In this chapter, as in the next one, we consider computation on text that can be done with the default, core installation of Python—without having any additional libraries or modules downloaded and installed. This chapter covers the simplest ways of dealing with text, using strings and slices. In the next chapter, the powerful concept of the regular expression is introduced. Later, in chapter 15, "Text III: Advanced Text Processing," the textual work is based on TextBlob, an additional library that provides several sophisticated capabilities for text processing and lexical work.

Those complex capabilities are great, but it's possible to get a very good start and learn about the computational manipulation of text simply using the standard, built-in capabilities of Python, focusing on just strings and slices. Knowing about the basics is important for working with additional libraries and frameworks, too.

[8.2]   Strings, Indexing, Slicing

We've seen already that one of the built-in data types in Python, and a very useful one, is the string. The *string* is a sequence of characters, indicated with surrounding quotation marks—either single or double, as long as both starting and ending quotation marks are the same. Examples of strings include `'hello world'` and `'2112'`. It's fine for a string to have no characters in it—`''`—and in that case it is known as the *null string*. (We saw this string in 7.5, Converting a Number to Its Sign.) Check out the lengths of the strings just mentioned using the `len()` function:

```
len('hello world')↵
```

```
len('2112')↵
```

```
len('')↵
```

These sorts of string values can also be assigned to variables, of course:

```
hi = 'hello world'↵

len(hi)↵
```

In Python, you can use double quotes instead of single quotes, as I mentioned:

```
len("hello world")↵
```

This is convenient in some cases—for instance, if you want to define this string: "I can't go on." If you use double quotes on the outside, it's allowable to have single quotes (apostrophes) inside them. There are ways to include the apostrophe/single quote without using double quotes, however. A Python programmer is never forced to use one or the other. I use single quotes to surround strings, mainly because they are slightly easier to type: it's not necessary to hold down the Shift key.

Python is offering some flexibility here, providing an option that makes it a bit easier to program. You might consider that both American-style and British-style quotation, with different nesting of single and double quotation marks, is allowed. But this doesn't mean that Python is completely flexible and that *every* style of quotation will work. Python is still a formal language in which there are some acceptable ways to quote strings and other ways that aren't acceptable. You don't need to try this one, but if you were to bust out some *guillemets* and type something like this:

```
"bonjour monde"↵
```

you would get an error message rather than defining a string. It doesn't matter that this is a culturally and linguistically acceptable way to quote things; it isn't a way that works in Python. Python's flexibility is only that which is defined in the programming language.

Notice that strings are not numbers. They are sequences of characters. As discussed with regard to types, integers don't have a length in Python:

```
len(2112)↵
```

Here we tried to find the length of a number, an integer. But that just isn't a sensible operation according to Python. The len() function tells us how many elements there are in a sequence, such as a string, which is a sequence of characters. It isn't meant to tell us anything about numbers—and indeed, 2112, the number that is 1056 * 2, does not have a length. But if we treat the four digits of the number as a text, we can find the length of the string '2112' using the length() function. The integer 2112 does not have a length in characters, but the sequence of characters '2112', the title of an album by Rush, does.

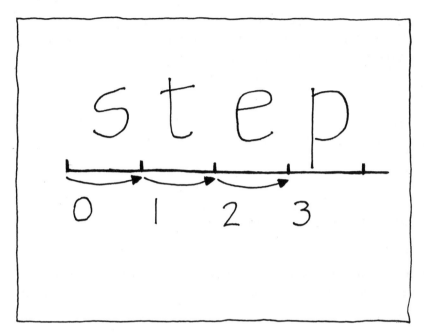

[Figure 8-1]
The *index* of the first element of a list is 0. That's how many steps are needed to get there from the starting point. Here, 's' has index 0, 't' has index 1, 'e' has index 2, and 'p' has index 3.

Strings can be thousands, or tens of thousands, of characters long, or even longer. An entire file can easily be read into a string; this is commonly done, and we'll do this later as we analyze and manipulate larger texts.

It's often the case that we want to examine or work with a *part* of a string. For instance, we might want to check the first letter to see if it is capitalized. Or we might be interested in how the first half of a string differs from the last half. In many cases we could accomplish this by finding a *substring*, although in Python it is often more convenient to take a *slice*. The details of the differences between these two can be left for later; for now, you should know that you'll find a slice very appealing, even if you don't like pizza. In Python, we can also easily access one character of a string using its *index*—its position in the string.

The first character of a string is number zero: the index of it is 0. When you are counting limes at a market, you probably don't begin counting "zero, one, two . . .". But there are sound reasons to begin at zero when it comes to the characters in a string. Character number zero is the character that is *zero* away from the beginning of the string. The index, in this case, is indicating how far you need to go (starting at the beginning of the string) to get to the character in question. This can be thought of as an offset. How far to go to get to the first character? Zero steps! You're already there. How far to get to the second character? One step gets you there. So that second character, one place away from the beginning, is character one, with index 1. Try this to see:

```
'hello'[0]↵
```

```
'hello'[1]↵
```

```
'hello'[2]↵
```

```
'hello'[3]↵
```

```
'hello'[4]↵
```

While we're at it, we might as well try:

```
'hello'[5]↵
```

There are only five characters—numbered zero through four—so it shouldn't be a surprise that asking for character number five produces an error. Does the error message make sense?

[8.3]   Selecting a Slice

To pick out one character, it's easy to just use the index. We can get any number of characters by taking a *slice*. For instance, the first three characters of a string correspond to the slice beginning at 0 and ending at (going up to, but not including) 3:

```
'hello world'[0:3]↵
```

```
'2112'[0:3]↵
```

```
''[0:3]↵
```

Now, what happened with that last example? The null string, which has no characters, of course doesn't have characters zero, one, and two. Shouldn't the Python interpreter have given us an error?

Actually, slicing a string, unlike obtaining a character by index, is a *forgiving* operation. Python will provide as many characters as it can, but if it runs out, or if there are no characters in the specified range, whatever is found in the slice will be returned—or the null string will be returned, if there's nothing there at all—and no error will be produced.

Notice that it's possible to get a string's first character by slicing:

```
'hello'[0:1]↵
```

Why would one do this, when it takes two fewer keystrokes to just get the character using its index? Try the following:

```
''[0]↵
```

```
''[0:1]↵
```

The second one doesn't produce an error and instead provides the null string as a result. A slice gives the specified characters if they're there and as many of those characters as possible if they aren't. In this case, it's a way of requesting the first character if there is a first character and '' otherwise.

Check out the slice with the first 500 characters of 'hello world', for instance:

```
'hello world'[0:500]↵
```

And check out the slice that holds characters 450 to 500 of that string:

```
'hello world'[450:500]↵
```

It's possible to leave out the first or last number when slicing a string, if you want to indicate "from the beginning" or "up to the end." To find the first three characters, for instance, one can also type:

```
'hello world'[:3]↵
```

And to find everything *except* the first character:

```
'hello world'[1:]↵
```

So, for instance, try:

```
'hello world'[:5]↵
```

```
'hello world'[6:]↵
```

Now, we can do something to just one part of a string. Let's say we want to just make the first letter of 'hello world' uppercase. To do this, we can apply the upper() method to just the first character, then append a slice that corresponds to the rest of the string. (A method is very much like a function, but attached to something, in this case a string—see the glossary for some details.) Give it a try:

```
greeting = 'hello world'↵
```

```
greeting[0].upper() + greeting[1:]↵
```

You can actually provide yet another number, beyond the starting and ending points, when slicing strings. This third number, or argument, is much less frequently used, but it's there for those who wish to use it, and it's the *step*. So if you want to slice up `'abcdefgh'` by starting at the beginning (character 0) and going to the end (up to, but not including, character 9), but skipping ahead each time by two, that is done as follows:

```
'abcdefgh'[0:9:2]↵
```

It turns out that both the first argument (starting point) and the second argument (end point) can be omitted, and if they are, the slicing will start at the very beginning and go to the very end. Because we can leave out the beginning and the end, the same thing just done can be accomplished as follows:

```
'abcdefgh'[::2]↵
```

Similarly, if you'd like to get all the even digits:

```
'0123456789'[::2]↵
```

This may seem like an esoteric way to slice a string. It certainly seems that way to me, even though this third number has its uses. I'm not asking you to internalize all of this, including the third *step* argument, right now. I'm just showing that there is more to the slice than the start and end point. A nice thing about Python is that you *can* slice strings just by using the first two arguments—and in many, many, cases, that will be enough. But if you want to take every other character, or every third character, for some reason, you can add the *step* argument to what you already know and allow your program to do more. There are not only optional arguments, but also ones you don't need to worry about until the need arises.

Finally, there is an easy way to get the last character, or the last several characters, of a string, without even determining the length of the string. The last character is at index `-1`, the second-to-last at `-2`, and so on. These negative numbers can be used in slicing.

```
'hello world'[-1]↵
```

```
'hello world'[-1:]↵
```

```
'hello world'[-5]↵
```

```
'hello world'[-5:]↵
```

[8.4]   Counting Double Letters

Before moving on, we can put our knowledge of slices to use. We can write a function that counts double letters in a string. Type in this string, all on one line:

```
wyatt = 'They flee from me that sometime did me seek / With naked foot, stalking in my chamber.'↵
```

These two lines aren't spelled this way in their original printing, but we'll use this version. And, although it doesn't matter for this specific input, we want to consider capital and lowercase letters to be the same, so that if our string begins with "Oolong" we will correctly identify the first two characters as a double letter. To accomplish this, just use the method that returns an all-lowercase version of a string, lower():

```
'HELLO World'.lower()↵

original = 'Burma Shave'

lowercase = original.lower()

lowercase↵

original↵
```

In typing the preceding code and observing the result, you should see that lower() does not change the value of the original string; it returns a new all-lowercase string. If we like, however, we can overwrite the original value with the lowercased string:

```
wyatt = wyatt.lower()

wyatt↵
```

Let's start our double-letter-detection process with an even simpler version of the rather simple function we're trying to write. Let's iterate through wyatt and simply find all the occurrences of a double *e*. To do this, our program should go through the string character by character, and at each point, it should consider the next two characters and check to see if the string containing them is 'ee'. To get going, we won't even write a function, although we will do that eventually. For now, we'll just hard-code our string, stored in the variable wyatt, into a loop.

Here's how to iterate through the string a character at a time:

```
for c in wyatt:

    print(c)↵
```

wyatt is a sequence (and specifically a string), and this code iterates through it one element (one character) at a time. The print() function shows us what value is contained in c at each step. Remember, it's just displaying this value for us, not returning each value.

Here is some code that does the same thing; type it in and run it to confirm that it does:

```
for i in range(len(wyatt)):

    print(wyatt[i])↵
```

Let's modify this slightly to see what's happening:

```
for i in range(len(wyatt)):

    print(i, wyatt[i])↵
```

We have i, an integer, before a comma and `wyatt[i]`, a character, after it. You can print any sequence of values by separating them with commas, which cause a space to be printed between them. For a quick peek at what values are, this works well.

Having run this, you can more clearly see that the extra variable, i, which has been introduced in this modified code, is being used as an index. It starts at 0 and is incremented until it reaches 85, because that sequence is what is produced by the range operator, as you can see:

```
list(range(len(wyatt)))↵
```

The string wyatt is eighty-six characters long:

```
len(wyatt)↵
```

So `range(len(wyatt))` counts out the eighty-six characters, starting at the first one at offset 0 and ending at the last one at offset 85. `wyatt[i]` then retrieves the appropriate character from the string.

That's great—we had something simple that worked, and now we exchanged it for something more complex that works. There is a point to making the code more complex, though. Our first, simple example could only access one character at a time, and we are interested in checking two characters at a time. The complexity we added allows us to do that; it's now simple to take two-character *slices* of the string rather than just examining a character at a time.

```
for i in range(len(wyatt)):

    print (wyatt[i:i+2])↵
```

The first problem (counting every case in which a double *e* is found) is almost solved; we can now identify each pair of letters. We just need to have the program count how many of them are `'ee'`. Let's add a counter to keep track of what we find, and let's add one to it whenever we find a double *e*.

```
pairs = 0

for i in range(len(wyatt)):
```

```
if wyatt[i:i+2] == 'ee':

    pairs = pairs + 1↵
```

After running this code, check the value of `pairs` to ensure that it's a sensible one:

`pairs↵`

Now we have code (not yet bundled into a function) that will count how many times there are two consecutive *e*'s. But we are looking to count all double letters, including the double *o* in *foot*. We need to replace the condition `wyatt[i:i+2] == 'ee'` with an appropriate condition. What we really want to test is whether the *ith* character, character number *i*, is the same as character number *i+1*. Let's try the most straightforward way of testing that:

```
pairs = 0

for i in range(len(wyatt)):

    if wyatt[i] == wyatt[i+1]:

        pairs = pairs + 1↵
```

If you type this in as it is printed, you'll get the same error I got. But check `pairs` again to see what its value is:

`pairs↵`

Interesting: `pairs` has the correct value, but this code resulted in a "string index out of range" error. So the program must have tried to access a character that doesn't exist. Because it seems to have done its work first, it may have tried to access a character that is beyond the last character in the string. Recall that `i` starts at 0 and goes up to 85; the final character in the string is number 85. But our program not only refers to `wyatt[i]`, it also refers to `wyatt[i+1]`. So at the very last step, it's trying to access the character at offset 85 *and* the character at offset 86 of the example. However, there is no character 86 steps away from the first one. Try it and see:

`wyatt[86]↵`

That's what is causing the error.

There are a few ways to fix the problem with the final comparison. Essentially, the program shouldn't be attempting to make this comparison. It's asking, "Is this string's *final* character the same as the character after it?" That question makes no sense. All the other comparisons make sense, but not this last one. So, let's just make one fewer comparison. Instead of having `len(wyatt)` determine the limit of our iteration, we will use `len(wyatt)-1`.

```
pairs = 0

for i in range(len(wyatt)-1):

    if wyatt[i] == wyatt[i+1]:

        pairs = pairs + 1↵
```

Now there's no error produced, and the result that remains in pairs is correct, as it was before.

[Exercise 8-1]   A Function to Count Double Letters

Bundle this code into a function, twin(), that returns the correct answer. It's best to type everything again, line by line, ensuring that you define a particular argument (the string to be checked) and use that in the body of the function. After writing this code, test your function. Test it on 'Oolong', for instance, to see if you are correctly considering the lowercase version of the string. The four lines of code given previously are not sufficient to deal with that string properly. You will need to recall, or review, how the string being checked was converted to lowercase. Then, you should include this conversion in your function.

[8.5]   Strings and Their Length

Strings are very useful for representing textual data or generating new data, and indexing into them and slicing them is only the beginning. We'll start this section with a short prose paragraph, Article 1 of the Universal Declaration of Human Rights, the English text. The text of this document can be found online at:

un.org/en/universal-declaration-human-rights/

You can find Article 1 just after the Preamble, or you can type in the two sentences yourself. Here's what to enter, all on a single line:

```
text = 'All human beings are born free and equal in dignity and rights. They are
endowed with reason and conscience and should act towards one another in a spirit
of brotherhood.'↵
```

After you've done that, enter just the variable name text in the interpreter. You should get all of this text back, quoted, because this string currently has the value of the variable text.

A simple question to ask is how long this string is—how many characters are in it. We can figure this out by simply checking the length of the string, using the len() function. Take a guess and then try:

```
len(text)↵
```

The result should be 170, assuming you didn't add spaces anywhere or mistype or mispaste the text. Some people are concerned with character counts; it's not a bad measure of how much text is there. (I have occasionally written book reviews that are exactly 1,024 characters long, and some conferences ask for abstracts and biographies of no more than a certain number of characters.) But we could certainly enrich our understanding of a text if we were able to consider the word level as well. So, let's see how we can compute in a per-word manner.

[8.6] Splitting a Text into Words: First Attempt

Strings provide a very useful method for this particular case, looking at text word by word. This method, split(), divides a string into a list of strings, breaking it apart whenever the specified string (the one given as an argument) is found. Try this and see:

```
text.split(' ')↵
```

The result is our string, but split into a list of words. The punctuation that is next to a word stays with it. Type text again to see that this method *returns* a list with the split-up string as elements but does not *change* the value of text itself or the value of whatever string it is invoked on. Check this yourself by looking at the current value of text:

```
text↵
```

Providing a new result while keeping the original value intact can be very useful. There are plenty of cases in which we would want to count the words in a string without destroying, or overwriting, the original string. On the other hand, methods that do modify their objects, when the programmer would prefer that they do not, are not deeply problematic, as long as the programmer remembers (or figures out) in which cases a change is being made.

We can check the length of the split-up string, text.split(' '), as a way of determining how many words are in the text:

```
len(text.split(' '))↵
```

To see that you can split a string using *any string* as the point of division, try using the word 'and' as the argument—after guessing what will happen:

```
text.split('and')↵
```

Using `split(' ')` is a first approximation to counting words; it works fine on certain well-behaved and nicely typeset texts. What if we had a text that had more than one space between some of the words? For instance, let's put three spaces in the middle of this string:

```
hi = 'hello   world'

hi.split(' ')↵
```

The result makes a certain amount of sense, but it is a bit unusual—and it is probably not what we want for word-counting purposes. The string is split into `'hello'`, the null string, the null string again, and finally `'world'`. The three spaces in the middle were used to divide this string into four parts.

Note that this result is exactly the same as:

```
'hello   world'.split(' ')↵
```

Yes, this line produces exactly the same result as when the two previous lines are both run; do try it out to be sure. The only difference is that in the previous two lines, the string `'hello   world'` is given a label, or equivalent, assigned to a variable, `hi`. This means that later on, it's possible to refer to `hi` rather than writing out `'hello   world'`. And if we transform what we have labeled `hi` in some way—assigning a new value to that variable—that variable will refer to the transformation.

There's still the problem of an improperly split text. The way to deal with this is to use `split()` without giving it an argument at all. If no argument is provided—that is, if nothing is in the parentheses—`split()` uses *any sequence of whitespace* that it finds to divide the string: not only one or more spaces, but also any run of spaces, tabs, and newlines. This means that when it encounters three spaces in a row, it considers that whole sequence of spaces to be the divider. This gives us a more sensible means of dividing a text into words. Try it and see:

```
hi.split()↵
```

Notice that this line is making use of the convenience that the variable `hi` provides us. We don't need to type (or copy and paste) `'hello   world'` again; we can simply type these two characters. That's of course not the only reason to use variables, but it is one.

Because there are no runs of more than one whitespace character—no sequence of two or more spaces, for instance—in our string from Article 1 of the Universal Declaration of Human Rights, the result of `text.split()` is the same as the result of `text.split(' ')`.

[8.7] Working across Strings: Joining, Sorting

Let's start this section by assigning a short list of people's names to a variable called names:

```
names = ['Bob', 'Carol', 'Ted', 'Alice']↵
```

Because we can split a string apart into a list of strings, it should not be too surprising that we can join a list of strings together into a single string. We simply use the join() method, like so:

```
' & '.join(names)↵
```

Note the spaces around the &. The only odd thing about this is that join() is not a method of the list. It's a method of the *string* that is used to do the joining—the conjunction, if you will. The *argument* to join, included in the parentheses, is the list whose elements (all of them strings) are to be joined together. Try it the other way around to see that it won't work:

```
names.join(' & ')↵
```

The error message (AttributeError: 'list' object has no attribute 'join') may not be crystal clear, but you should be able to see that it's at least consistent with the discussion so far. A list doesn't have a join() attribute (specifically, a method). Strings *do* have this method, and so we need to invoke join() on a string.

Check to see that just as split() leaves the string it is invoked on intact, join() also leaves its list argument intact:

```
names↵
```

Now try this again:

```
' & '.join(names)↵
```

Invoking ' & '.join(names) results in the title of Paul Mazursky's 1969 film. But we could use a different conjunction, too. Try:

```
' and/or '.join(names)↵
```

There are many ways to format output to make it more legible. At this point, I'll note that join() can be used for that purpose. In Python, a newline character is represented within a string as \n. If this newline is used as the conjunction, the list is effectively typeset with one item per line:

```
'\n'.join(names)↵
```

The result here might be even less exciting than you expected. The Python interpreter shows you \n rather than actually producing a newline. But try printing this to see what the result will look like:

```
print('\n'.join(names))↵
```

As you probably expect, you can include other text in addition to a newline. Type this in, think about what it will do, make a guess about the output you expect, and press Shift-Enter:

```
print(' exists.\n'.join(names))↵
```

Did you expect the result to claim that Bob, Carol, and Ted exist? But not that Alice does? Remember, join() is placing a sort of *conjunction* between strings in a list. It isn't adding anything to the end (or beginning) of the list. Nor is it doing something before or after every element. But if you want to add something to the end of any string, you can do it with string addition, using +. The entire expression ' exists.\n'. join(names) returns a string, so simply put a + after it and then conclude with the string necessary to assert everyone's existence. Figure out what this concretely means in Python and give it a try.

If you followed the suggestion, you will have found one of several ways to go through a list of strings and print each string with ' exists.' appended to the end of each one. We also added a newline to the end of every string except the last one. I consider that it's not wrong (it does actually work), but this code is also not optimal. The purpose of join() is to add the same conjunction between each string in a list. In this last example, we really didn't want to do that; we wanted to add something to end of each string. But we got to that point in a quasi-exploratory way, first using the newline as a conjunction and then seeing that we could make each name the beginning of a short sentence.

Programming is full of different ways to do the same task, regardless of programming language. In exploratory programming, I find it best to first get a program *working*, functioning in the intended way, and then (in at least some cases) worry about using the *best* method, or at least a better and reasonably good method, by changing the code in a process of refactoring.

The current method, even though it can be accomplished in a single line of code, isn't actually the best one. For one thing, the string ' exists.' is repeated in two places, as the conjunction and as the string added at the end. If this were part of a longer program, I might decide at some point that instead of using ' exists.', I would like to say, ' just simply is!'. But I'm inviting a mistake, because in the current code

I am able to change ' exists.' in one place and accidentally leave the original text unrevised in another place. To repeat myself again: don't repeat yourself.

Does it not seem very likely that a careful artist or humanist would make this sort of mistake? Maybe not. But consider that if one line of code has the possibility for a mistake of this sort—an unnecessary possibility—then one hundred lines of code, which is really a rather short program when one is considering a reasonable research or art project, could introduce the possibility of one hundred different types of mistakes. In a quick sketch, it may not be worth it to *refactor* and produce new code that does the same thing in a better way. But when more than a few lines are being written, and there's a chance to improve and foreclose the possibility for future mistakes, I recommend it.

[8.8]   Each Word without Joining

As a first example of refactoring, here is a better way to produce the same text:

```
for person in names:

    print(person + ' exists.')⏎
```

This is two lines long, and thus less concise, but it certainly is very straightforward. Instead of using the newfangled join(), this code simply iterates through the list, and for each string in it (each person) it prints that string with ' exists.' added to it. There is no need for explicit newlines (each represented as \n) because print produces a newline each time after it is done outputting each string.

Now, join() is not entirely useless simply because we eliminated it in this particular case. There are just different situations in which it is best to use it.

While we have names handy, let's see how we can sort lists. To start with, let's use a way of sorting this list that is similar to split() and join() in that it *returns* the value we're interested in—the sorted list—and does not change the value of the variable that we start with. In other words, we want to get a new sorted list without modifying names. Here's how to do it, using the sorted() function:

```
sorted(names)⏎
```

Check to see that after this function has done its work, names is unchanged:

```
names⏎
```

Very nice. There is also another way to sort sequences, and this way sorts them *in place*. In other words, the original list itself is sorted, and no new list is produced. This can be efficient for very long lists. In extreme cases, when the list occupies more than half of the computer's memory, it may be possible to sort the list in place and impossible to return a sorted copy of it because there is no room in memory for the copy. When there is truly no need to retain the original, unsorted list, sorting in place is a good idea. It can be done in this case with `names.sort()`. Don't take my word for it. Try it out!

By now we have covered both the essential, general programming fundamentals and some of the ways that we can work with a specific medium, text, using strings, character indexes, slices, splitting, joining, and sorting. Not only has iteration been explained, with different examples of going through lists and performing calculations—but we have also seen how to iterate through a string and examine slices of it to determine double letters. This is a critical point in the book where you should be able to put together the fundamentals of programming and the basic ways that Python allows you to work on strings to accomplish simple text analysis, manipulation, and generation. Once you do so, you will be ready to move to more complex work with text, to work with images, and to the other topics of the book. To ensure that you understand the essentials so far and can proceed, at this point I present not a free project but several exercises.

[Exercise 8-2]   Same Last Character

Write `same_last()`, a function that accepts two strings as arguments and returns `True` if they have the same last letter, `False` otherwise. For this exercise, you can assume that both of the strings are at least one letter long; it does not matter what happens (the program could crash, etc.) if one or both of the strings is the null string.

After you write a function that works, take a look at the body of the function you wrote—the part after the `def` line with the colon at the end of it. How long is that code block? If you have more than a single line of code there, you should refactor. You can solve this problem with a function body that consists of only one line.

[Exercise 8-3]   Counting Spaces

Write `count_spaces()`, a function that accepts a string as an argument and returns the number of spaces in the string. Use iteration to determine this.

If you can think of more than one way to accomplish this, write `count_spaces_2()` and go on to write `count_spaces_3()` and beyond if you like, showing the alternatives. To accomplish the basic, initial `count_spaces()` function, no special knowledge of Python is needed beyond what has already been covered in this book. If you can't determine how to solve the problem with the techniques explained, you need to review the book and learn the fundamentals, determining how iteration can work in this case. Then, if you like, you can seek alternative ways.

[Exercise 8-4]   Counting Nonspaces

Write count_nonspaces(), a function that returns the number of characters in a string that are *not* spaces. Try figuring this out using iteration, with reference to the problem just solved. Once you have solved the problem this way, see if you can determine how to do this in a single line (not counting the line beginning with def) by having count_ nonspaces() call your count_spaces() function.

[Exercise 8-5]   Determining Initials

Write initials(), a function that takes a string containing any name (e.g., a personal or business name) and returns the initials. For instance, the values returned by the following function calls will be:

initials('International Business Machines')

'IBM'

initials('Georges Perec')

'GP'

You should be able to tell what type the *return value* (i.e., your result: the initials) should be. The function should work properly on names with any number of words, whatever the length of the words. You do not have to worry about special handling for cases where punctuation makes up its own "word" or where a word begins with a punctuation mark or where you know that a compound word should really provide two initials instead of one. Just include the first character of each part of the string separated by whitespace. For instance, as we have defined the function here, the following return values are actually correct:

initials('Country, Bluegrass, & Blues')

'CB&B'

initials('Vladimir "Pootie-Poot" Putin')

'V"P'

[Exercise 8-6]   Removing Vowels

Write devowel(), a function that accepts a string as an argument and returns the string without the vowels. For instance, given 'hello world', it will return 'hll wrld'. Just consider the five standard, full vowels for this exercise, neglecting poor y and w. We haven't discussed how to remove characters from a string, but you can still solve this problem. Develop a solution that involves building up a new string, leaving the vowels out as you do so.

[Exercise 8-7]   Reduplications

Write reduplication(), a function that accepts a string and returns True if the string consists of some sequence of characters (call it A) followed by the same sequence of characters, A. The function needs to return False otherwise. For instance, given 'hello world' it will return False but given 'worldworld' or 'aa' it will return True. Of course, for 'worldworldmoon' or any other reduplication with something stuck on the end of it, the answer is False. We will concern ourselves only with strict lexical reduplications for this exercise.

To figure this one out, you probably want to find the *midpoint* of a string, the place halfway along its length. But if you do this with standard division, using /, you'll get a number with a decimal point, which can't be used to refer to a point in a string. You either can use integer division, which is done with the operator //, or you can convert your answer to an integer with int().

[8.9]   Verifying Palindromes by Reversing

A *palindrome* is a text that reads the same forward and backward. Actually, any sequence of discrete tokens or objects—a genetic sequence, the digits of a number, a row of soft drinks arranged in a refrigerator rack—can be palindromic or not. To begin, though, consider words represented as strings.

We will focus on single words in which the unit of reversibility is the letter. Palindromes in this sense are *civic*, *racecar*, and *kayak*, for instance. There are other words that come close (in some sense) to being palindromes but aren't. The word *revere*, for instance, would be palindromic except for that *e* on the end. It isn't difficult for people to distinguish palindromic and nonpalindromic words, but we might want to do this for every word in a very large document or even a corpus of documents. We might be interested to know if authors such as Nabokov and Poe, who had a significant interest in palindromes, used more palindromic words than other authors. Or we might want to know if a particular language, as represented by a particular corpus, typically has a higher percentage of palindromic words than another, similarly represented.

Let's write a very simple function to return True if we have a one-word palindrome and False otherwise. Initially, we won't care what this function does if we give it something other than a single word. We also only expect it to work if the case of the letters is the same: *civic* is a palindrome, but *Civic* is not (sorry, Honda).

Now that you have learned the fundamentals of programming, you should be ready to try a task like palindrome validation on your own. Can you fill out the following partly completed template so that it gives the right answers?

```
def palindromic(text): ↵
```

———————

Take a few minutes, at least, to try to come up with an answer, based on what I've mentioned about palindromes and what you know about programming. If it becomes too frustrating, you can just read on, but give yourself a chance to develop a function and see how you approach the problem.

There are several ways to go about writing such a function, and we'll go through several of them. To begin, we'll develop a function that accepts a string, that has a string argument, but internally tests whether or not *lists* are palindromic. Lists and strings are both sequences, but instead of testing the string 'civic' we'll test that sequence of characters stored in a different way, as a list: ['c', 'i', 'v', 'i', 'c'].

We can build lists of characters from strings easily, using casting:

```
list('rotor') ↵
```

And of course, these lists can be labeled (assigned to variables):

```
to_test = list('rotor') ↵
```

There are a few good ways to test for palindromicity. When we're testing lists, it's very useful to know that there is a built-in way to reverse them:

```
test = ['a', 'b', 'c']

test.reverse()

test ↵
```

In the second line, the list labeled test is reversed *in place*. Just as the sort() method sorts a list in place, this reverse() method reverses it in place, so that the original, unreversed list does not remain in memory; only the new result is there. How did you know that reverse() was available as a method of lists? In this case, you learned it by reading a book that introduces programming—this book. You could have also learned about this by consulting documentation or tutorials online, by asking someone else (in person or online, in a forum), or by using Jupyter Notebook to inquire about what methods are provided by lists, as explained in 10.3, Generating Very Simple Images.

Now, why did we bother using the list representation, when strings are much more obvious ways of representing text? Lists are being used here because they have a reverse() method and we want to reverse the data we have. Strings don't have such a method. The designers of Python must have figured that if people were trying to have their program reverse a string, they were probably making a mistake rather than undertaking some serious activity such as palindrome checking. Check for yourself to see that the reverse() method isn't available for strings:

```
text = 'abc'

text.reverse()⏎
```

Let's try manipulating a word to see if it's the same forward and backward:

```
word = 'hierarchy'

backlist = list(word)⏎
```

What do you expect to be stored in the variable backlist? What will the type of data be, and what specific data will be stored there? Formulate an answer and check it:

```
backlist⏎
```

Now—because backlist is the right *type* of object—let's reverse it:

```
backlist.reverse()

backlist⏎
```

The original, unreversed list is gone, replaced by the reversed list. But remember that the variable word still has the original string, and that string can once again be made into a list:

```
word

list(word)⏎
```

You should be able to see that there's an expression that yields the word as a list, a character at a time, from beginning to end. And there's another list in memory, labeled by a variable, that represents the word as a list, a character at a time, from *end* to *beginning*. The question of whether the word is a palindrome is simply the question of whether these two lists are equal, which can be tested with ==. Type in the expression, which will be True for a palindromic word and otherwise (as in the current case) will be False.

You have just done all the work that a function needs to do to determine whether a word is palindromic or not. You haven't written the function yet (I presume), but you have tried out all the code it will use in the interpreter. Now, put it together as follows, with this as the first line:

```
def pal(word):
```

This line indicates that you are defining the function pal(), which accepts one argument, word. You should use three additional lines to implement the function that I have tried to lead you to discover in the previous text. The three lines should do the following:

1. Define backlist as the list containing the characters in word.
2. Reverse backlist.
3. Check to see whether the list of characters in word is the same as backlist; return the result.

Only four lines (including the standard first line of a function definition) are needed. There is no need to use the keyword if statement—that is, to use the conditional statement—to return True in one case and False in other. Your expression to check for equality *already* produces True or False as is appropriate. Just return that value directly. If you're totally puzzled about how to do this, it's okay to first develop the function pal() using a conditional statement and then refactor to remove it.

After developing this, enhance it so that it works regardless of typographical case—so that 'Civic' returns True, for instance. To accomplish this, just use the lower() method that was introduced in 8.4, Counting Double Letters. Modify pal() so that it compares the characters in the *lowercased* word with the reversed characters in the *lowercased* word.

The method developed so far has some nice qualities, but there's also something irksome about it. The use of lower() to lowercase the string doesn't seem to be a bad idea, if we want to ignore case. But we then convert our string to a list, reverse the list, and convert the string again so we can compare the two lists. On the one hand, this represents something fundamental about programming languages: they almost always have different types of data (strings, lists), one can convert among these types, and certain things can be done using some types that cannot be done with others. On the other hand, needing to convert back and forth in this way can be rather annoying.

Even though there is no reverse() method for strings, there is a way to easily (and very quickly) reverse them. Here's how to start with a string containing the alphabet and get a reversed string:

```
'abcdefghijklmnopqrstuvwxyz'[::-1]↵
```

Alternately:

```
alpha = 'abcdefghijklmnopqrstuvwxyz'
```

```
alphaback = alpha[::-1]
```

```
alpha↵
```

```
alphaback↵
```

From typing this in, you can see that this method doesn't change the original string; it produces a new string. But what exactly is happening here? As discussed in 8.2, Strings, Indexing, Slicing, this operation is a way of *slicing* the string. Specifically, it starts at the beginning (there is no first number), goes to the end (there is no second number), and proceeds using a *step* of -1—yes, *negative one*—which corresponds to moving one character back *each step*. The slice is therefore done one character at a time, but backward.

Is this an obvious way to reverse a string? If you have a python wrapped around your brain stem, perhaps it is. It seems most obvious to many newcomers, and to several programmers who are familiar with the English language, to try to use the `reverse()` method instead. But this slicing technique works, is a very fast operation, and can be expressed compactly in code. Refactor your `pal()` function using this method of producing a slice with step -1. You should be able to write a function that consists of the function definition (beginning with `def`) and only one additional line.

[8.10]   Verifying Palindromes with Iteration and Recursion

Having figured out how to solve this problem, we'll abandon the extremely efficient and concise method of comparing a string to its reverse. We don't desperately need to ship a palindrome verifier by some deadline so that Apple's App Store can choose whether or not to approve it. We're doing these exercises to learn about programming and how to solve problems. We could try to solve three different problems using three different approaches, but instead we will hold the problem constant and see how different methods can be used to solve it. Being able to approach a program in different ways is one foundation for exploratory programming.

Next, we will use palindrome checking as a problem that can be solved through *iteration.* And after we're done with that, we can use this same problem to understand how the same task can be accomplished with *recursion.* These two approaches were introduced in 7.6, The Factorial. There you found that either one could be used to compute the same result. In this return to iteration and recursion, we'll see that sometimes one or the other method is actually more advantageous. This is why it often helps to be able to think about computing in more than one way.

The iterative version of palindrome-checking, as described here, involves going character by character through the string, forward, and at the same time going character by character through the string, backward, comparing each character. The first comparison we want to make is between the first character and the last character. The index (or offset) of the first character is 0, while we can use index (or offset) −1 to access the last character:

```
original = 'Civic'
string = original.lower()
string[0]↵
string[-1]↵
string[0]  == string[-1]↵
```

Then, the second comparison we want to make is between 1 and −2. The third is between 2 and −3, and so on. If we like, we can make as many comparisons as there are characters in the string. Let's use the variable right to hold the *right-side index* that starts with −1 and decreases:

```
right = -1
reverses = True
for letter in string:
    reverses = (reverses and (letter == string[right]))
    right = right - 1↵
```

Initially, the program assigns the variable right to have the value −1; that's the special index, indicating *last character*, that we start with as the program goes backward through the string. Also, reverses is set to True. We assume the string is palindromic until we find out otherwise. As the program iterates through string, it compares the current character (from our forward iteration) with string[right] (from the other side of the string). The expression reverses = (reverses and (letter == string[right])) updates reverses to remain True (if the characters are equal) or to become False (otherwise). This essentially creates a long logical conjunction: "(the first and last characters are the same) and (the second and second-to-last characters are the same) and . . .". If even one of these is False, the whole expression is False. If all are True, we have a palindrome.

This is a little hard to understand, but it could have been worse. If I had written:

```
reverses = reverses and letter == string[right]
```

Instead of:

```
reverses = (reverses and (letter == string[right]))
```

The code would still have run properly, but it would have been harder (particularly for a new programmer) to see what was going on. The way I've written it, it's clearer that all of that stuff on the right is being assigned to reverses. Specifically, we're assigning it the value of the conjunction of reverses itself (so that if it becomes false, it says false) *and* this other equation, which will be true as long as those opposite characters are true. We're not checking to see if reverses and letter is equal to string[right]. The parentheses make that clear. While they aren't needed for the program to run and work correctly, parenthesizing can still be a help to those seeking to understand the program or to develop the program in the first place.

The only other part of the code, the final line, decrements right (i.e., subtracts one from this variable) so that it remains the *backward index* or *starting-from-the-right index* throughout the process. Try this code out to see that it works. To check out what is being compared at each step, modify it to add a print statement:

```
right = -1
reverses = True
for letter in string:
    reverses = (reverses and (letter == string[right]))
    print(letter + string[right] + ': ' + str(letter == string[right]))
    right = right - 1
```

As soon as you have this working, bundle this up into a function, pali(), which accepts a string as an argument and returns whether or not it is a palindrome. This will encapsulate your code so that it can be easily reused. Apply your pali() function to both palindromes and nonpalindromes. You should be able to trace through the process and see how the iterative approach works. Of course, you also need to make sure it's returning the appropriate value for strings of different sorts.

If you contemplate this for a while, you should see that this code makes unnecessary comparisons. This code compares the first half of the string to the second half of the string. But then it continues to compare the second half of the string to the first half, which is just repeating the work already done. You can, if you like, fix the program so it goes only through half of the string. But you could also take a different approach—you could simply not care about the extra work the program is doing. Yes, it means that the process will take twice as long, which could possibly be bothersome in some circumstances. But this is a sound method of checking palindromicity, and in attempting to fix it, you might introduce a bug. As with any time you have a choice to refactor, you should realize that it is a choice, not a requirement.

Why would we want the (relative) mess of code above, packaged into a function, instead of something very simple and elegant? For instance:

```
def pal(text):

    return text == text[::-1]
```

Or, if you like, a similar function with the method lower() applied on both sides of the test for equality? Well, one situation in which we might want the longer function is when we are learning to program and trying to understand different ways of tackling problems. But there is also a practical advantage to the iterative code, one related to the idea of sorting in place. Consider a truly enormous string that is being tested for palindromicity. Let's have this imaginary string be so large that it takes up three-quarters of the memory in our computer. If we use the code that makes a reversed copy of the string, that reversed copy won't fit in memory because the string and its reverse need 150 percent of the available memory. (True, we might use our hard disk to provide virtual memory, but whatever memory we have is ultimately limited in some way, and we can exhaust it.) The code to compare a string to its reverse won't work in this situation. But the iterative code will work in this case. It's a small program (a program which itself doesn't take up much memory) and simply compares the string two characters at a time without needing to make a copy. So there is this concrete reason to learn about both approaches.

There's also a more directly practical reason, which matters even for small strings. We might want to not only detect whether a string is a palindrome or not, but also to find out, if it isn't, where the first (or outermost) mismatch is. The method we used previously isn't going to easily allow that, but a different method will.

Instead of just slightly refactoring the iterative palindrome checker, let's write it entirely over again. Our first checker, pal(), implements this sort of idea: "A palindrome is a string that is the same as its reverse." The second checker, pali(), implements a different concept: "A palindrome is a string whose first character is equal to its last character, whose second character is equal to its second-to-last character, and so on." Now we'll implement palr(), a *recursive* palindrome checker, whose idea is this: "A palindrome is a string such that if you take off the first and last character, they are the same, and what remains is a palindrome." Recall that the first letter of our string is string[0] and the last one is string[-1]. You can determine the slice we want to take that has everything else ("what's left") in it, which is in between these two. Try

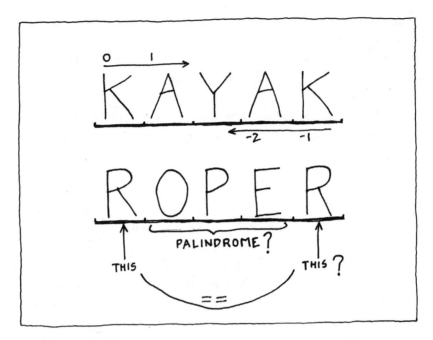

[Figure 8-2]

There are iterative and recursive ways to see if a string is a palindrome. "KAYAK" is being checked iteratively. First, is the first character (index 0) is the same as the last (index -1)? Yes, so is the second character (index 1) the same as the second-to-last (index -2)? And so on. "ROPER" shows the *recursive* way. Are the first and last letters the same? Yes, both are 'R.' Next, if anything more than a single character is left, use the recursive checker again on it, beginning stage 2. In this case, "OPE" is tested at the beginning of the second test, and 'O' is not the same as 'E,' so, no palindrome.

out some different slices in the Python interpreter to see what gives you all but the first and last character. You can use a string that you type in directly, such as `'hello'`, to do this if you like.

This recursive idea isn't yet precise. As we just described it, it will never "bottom out" by reaching a case that we know is palindromic (or not). We also need to add something else: "Any string of one character or zero characters is a palindrome." Here is the full implementation of this recursive idea, which includes checking a version of the string that has been converted to all lowercase:

```
def palr(text):

    lowercase = text.lower()

    if len(lowercase) <= 1:

        return True
```

```
    else:

        if lowercase[0]  == lowercase[-1]:

            return palr(lowercase[1:-1])

        else:

            return False↵
```

Remember that we conceptually defined a palindrome in terms of the concept of a palindrome. So it shouldn't be too big of a surprise that our function `palr()` calls itself.

Here's a question, similar to one that was posed before: Could `palr()` call `pal()` or `pali()` instead? Why or why not? Modify `palr()` so that it calls first `pal()` and then `pali()` to see if you are right. Is `palr()` calling `pal()` an example of recursion? If you aren't sure, read the discussion of recursion again and check the definition of it in the glossary. This type of mash-up isn't an example of recursion, nor would it be sensible programming practice, but if you are able to think through this, you'll understand a different important concept: how the interface to a function provides one sort of abstraction.

This is a great point at which to review the recursive function introduced in 7.7, "Double, Double" Again. It should now be easier to see how `doubler()` operates with reference to `palr()`. Trying these programs out should help you see that even if the idea seems somewhat odd at first, there is nothing mystical or metafictional about a function that calls itself. Recursion is just another approach to doing work with computation.

The palindrome checkers developed by this point will work for single words of most sorts. There are, however, more complexities to determining whether a given string is a palindrome. It's not a major concern in English, but in general, diacritical marks are usually ignored when determining whether a string reads the same forward and backward, so that *e* and *é* and *è* are all considered the same letter. (Of course, this does come up in English, as when you are working on your résumé in a café, perhaps in a naïve manner.) Also, in general, it's not just individual words, but also longer strings, that can be palindromes or not. For palindromic purposes, punctuation marks and spaces are generally ignored; numerals usually are not. So a more refined definition of the palindrome is that it is *a sequence of characters such that the letters and numerals in it are the same forward and backward*. And finally, the discussion so far has been about letter-unit palindromes, but even sticking to the topic of language, there are texts and utterances that are palindromic by syllable, by word, by line, by sentence, and so on.

If you decide that certain characters, such as punctuation, whitespace, and so on, just don't count when it comes to determining the palindromicity of a string, you may find that modifying the code to account for this is easier with some programs and harder with others. In this particular case, I find that the iterative checker is a pain to extend in this way, whereas the recursive checker is easier to work with. What seems to be the simplest way to do it, though, is to clean up the original string so that only characters and numerals remain, then check the cleaned-up string. And that is one of many tasks that is most easily accomplished with *regular expressions*, which are covered in the next chapter.

[8.11] Essential Concepts

[Concept 8-1]   Strings Can Be Examined and Manipulated

You should be able to access one character of a string using the index and be able to access any section of a string with a slice. Splitting a string and joining a list of strings should also be easy for you. Work at string manipulation further in Jupyter Notebook if anything is unclear.

[Concept 8-2]   Iterating over Strings, Accepting Strings, and Returning Strings Is Possible

It's also important to be able to integrate the ways you are accessing and dealing with strings with the fundamentals covered earlier. Strings are sequences; understand how to iterate through them. It should be obvious at this point how a function can accept a string (or two strings, for that matter) and how it can return a string if necessary. If the task is something such as counting spaces, however, the return value will need to be a different type. Which one?

[Concept 8-3]   Reversing, Iterating, and Recursing Are All Effective

The point of the palindrome-checking example was to show three different ways of approaching the same problem. Not only do these three methods each work, but they each have advantages. Reversing the text is simplest in terms of code length and program development; the iterative procedure uses the least memory; and finally, the recursive method is a clean decomposition of the problem into a single test and a smaller version of the problem. It's important to understand how these three solutions work because as your explorations grow more complex, it can be useful or even necessary to frame one particular task in different ways.

[9] Text II: Regular Expressions

[9.1]

Now that we have manipulated texts (by indexing individual characters, taking slices, splitting strings, and joining lists of strings), it's possible to consider the powerful, general approach of using regular expressions. Regular expressions are a formal language themselves, used in Python and also in many other contexts. While they do not do everything that a general-purpose programming language does, they are extremely useful for text processing.

[9.2] Introducing Regular Expressions

We very often want to transform texts or search for a particular sort of string within a text. There is a somewhat cryptic-looking but extremely useful, and really quite widespread, formalism for doing so: regular expressions.

Standard text editors—all but the most simplified ones—allow you to search and replace text using regular expressions. You can even search in a word processor using regular expressions. You can, additionally, search large files of yours (such as email records you keep locally) using regular expressions, either from the command line or using a text editor. This means that even if you never program after reading this book—an outcome I dread—you can still get a great deal of practical use out of regular expressions, improving the way you work with computers.

A regular expression, sometimes called a regex, can get extremely elaborate, but, as with most things related to programming, you can learn just a little about this area and get a lot of use out of that knowledge. The system of regular expressions allows you to define patterns, so instead of just searching for a particular string, you can easily search for whole categories of strings.

To begin, have a text editor ready, such as Atom or Geany (on GNU/Linux or any other platform), Brackets (Mac), or Notepad++ (Windows). Then, grab a large file online; this time, grab the Project Gutenberg plain text UTF-8 edition of *Pride and Prejudice*, which happens to be Project Gutenberg's most-downloaded book:

gutenberg.org/files/1342/1342-0.txt

Save the file to your desktop and open it in your text editor. Using the Find option, you can search this text using regular expressions. You may have to check a box that says something like Use Regular Expressions. There also may be an option that sets your search to be case-sensitive or case-insensitive; that is, your search treats capital and lowercase letters differently or treats them the same. So that our regular expressions work the same way as they will in Python, change your settings (if necessary) so that your searches will be case-sensitive.

A very simple task one can do with the Find dialog is to see how many times a particular string occurs. Use the Find dialog to see how many times the particular sequence of characters dance occurs in the text. Your answer should be sixty-four. Of course, you can search for an even shorter string if you like. Say, for instance, that you're curious about how often the lowercase letter *a* occurs in *Pride and Prejudice*. Well, just put that letter in the search box and click the button. Did you get 42,224? If you didn't, you may have downloaded a different *Pride and Prejudice* text file. It's also possible that you are doing a case-insensitive search. That is, you are searching for both *a* and *A* when you only want to find the lowercase letter. Use some sensitivity, and change your text editor settings so that your search is case-sensitive.

Using regular expressions, and in particular *patterns* that we define using the language of regular expressions, we can do many other sorts of searches. We found how many times the lowercase *a* occurs. Now, we might be interested in matching a small number of characters, but still more than just a single character. For instance, what if we would like to count the number of (canonical, full) vowels in a text? We could of course manually search for a, search for e, search for i, search for o, and search for u, and then add up the results with a pencil in our lab notebook. But then we would have to not only do the same sort of thing five times, but also do some arithmetic. It seems not only error-prone but also silly for us to do that when we're sitting right in front of a computer that can do such things for us.

Regular expressions allow us to define a character class of the five vowels. To define any character class, that part of our expression needs to start with [and end with]. In the current case, we can simply write: [aeiou]. Let's go ahead and have this character class explicitly include both capital and lowercase vowels: [aeiouAEIOU]. That wasn't very difficult, was it?

Be aware that your text editor may not use regular expressions by default. In Atom, the editor I use, there's a button that has ".*", and that can be clicked on or off. I need to click it on to search using regular expressions, just as I need to click on the "Aa" button to make my searches case-sensitive.

Try using this list expression, this last pattern, to determine the number of vowels in *Pride and Prejudice* (209,876 by my text editor's count). You may notice that there's a bit of computation going on as your text editor searches the text for all occurrences of the five vowels in both upper- and lowercase. It may take a moment to get a result. Consider, too, how much longer it would take for you to go through the text and determine this *without* a computer. If you have a corpus of literature, you can quickly determine which of the texts is most vocalic and which is least vocalic. Of course you'd want to divide by the number of letters overall—the expression that matches "any letter" is [a-z], or [a-zA-Z] should you want to include all uppercase and lowercase letters—and you would probably automate the process rather than running this search by hand on each file. For now, determine how many letters (overall) are in *Pride and Prejudice* and then what percentage of them are vowels.

At this point we don't have anything to compare against this result. Would a book of poetry from a similar time period have a larger or smaller percentage of vowels? Would another Jane Austen novel? Have literary works somehow become more or less vocalic over time? Indeed, is the question of which texts are more vocalic and which are less vocalic at all interesting?

To be clear, I'm not asserting that this question is a deep and compelling one. In order to address questions such as these, we would need to generalize what we've done in our text editor to a larger number of documents, probably writing some exploratory programs to help us. I want to leave these questions aside, actually. My point is not to interest you specifically in vocalic studies of this particular sort. Rather, my point is that if you want to poke around and find out something about questions such as these, you don't have to prepare a grant proposal or a thesis proposal or otherwise come up with an elaborate justification for taking a quick look. You also don't have to hire a programmer! With a small amount of programming knowledge, you can try out an experiment along these lines and see if it seems interesting or if it leads to something interesting.

[9.3] Counting Quotations in a Long Document

Let's now undertake an experiment which could be more interesting than vowel-counting. We will begin by counting all the one-line quotations that occur in our text of *Pride and Prejudice*. For our purposes, a one-line quotation is a string that begins with a double quotation mark, ends with a double quotation mark, and has any number of characters (even zero) in between. We really don't want to restrict ourselves to one-line quotations—it would be better to find quotations even if they run over several lines—but let's start with something simple and go from there.

When I developed the first edition of this book, the *Pride and Prejudice* text file that Project Gutenberg offered was an old-style ASCII text file, and used *inch marks*—straight double quotation marks, which are standard in ASCII but which no self-respecting typographer would use in a book. Well, this has changed in file `1342-0.txt`! We now have proper double quotation marks and Unicode (UTF-8) text. Note, then, that in the discussion that follows, a double quotation mark is *not* " but is either the opening double quotation mark, ", or the closing double quotation mark, ". Among other things, this means that you'll need to type these specific characters, or you'll need to copy and paste them from somewhere, such as the Wikipedia article "Quotation mark." I use an international keyboard layout and can quickly type ", ", or ", but whether you want to change your keyboard layout or not, you need to be ready to input the opening double quotation mark and the closing double quotation mark.

To get back to how regular expressions work, let me next explain that there is a special symbol used in regular expressions to mean, approximately, "any character." This is simply the period (or dot). To be precise, . is used to mean "any character except a newline." Because we are looking for one-line quotations, it's fine with us have a symbol such as . that will match everything except a newline.

Fortunately, in writing a regular expression, there is also an easy way to indicate something like "any number of characters." To say "any number of (whatever)" in a regular expression, you use a *quantifier*. The first one I'll introduce is *, indicating zero or more occurrences. Bear with me for a moment; it probably seems a little bit strange to match zero occurrences of something. But in combination with other parts of a pattern, it can be very useful to do.

Because the way to indicate "any character except a newline" is with a period or dot, and because the * means "zero or more of those," .* means "any number of characters except a newline." Now to match our one-line quotations, the pattern, the regular expression, that we can try is ".*" (a pattern four characters long that begins and ends with our friends, the opening and closing double quotation marks).

Note that the opening and closing quotation marks in this situation are *part of the pattern*. They aren't there to surround a two-character string, for instance; they are there because we're trying to match a pattern that begins with a quote mark and ends with a quote mark. Search for this pattern in the entire document and see how many times it occurs. If you have the same document that I do, the pattern will be found 729 times.

Now, it's not very interesting to search a single file and find the result. What does the result mean? If it were 22 or 8946 instead of 729—so what? But if we compare our results with those from another file, we might be able to learn something about how two texts relate.

So, go get this Project Gutenberg edition of *Leaves of Grass*:

gutenberg.org/files/1322/1322-0.txt

Again, it's the plain text UTF-8 version. I selected this book because it's also popular and is almost the same length (in characters) as *Pride and Prejudice*. It's not *exactly* the same length, and when we wish to compare how many of something (quotations, pronouns, etc.) are used in text A and text B, it really makes sense to normalize each value by dividing by the length of the text. But for now, let's just look at the overall number of results for each document.

If you search the entire document for dance, without requiring that the case matches, you'll see that it occurs thirty-two times in *Leaves of Grass*. If you search for the pattern "." (using the opening double-quotation mark and the closing double-quotation mark), you'll see only *six* matches.

This method doesn't work quite as well as it could. Our idea of a "short quotation," one that occurs within a line, is a rather provisional one. It might make some sense for lineated poetry, but less sense for a novel, because the way a particular compositor has broken the lines will influence what a short quotation is in that case. Also, we're not even counting short quotations properly. Take a look at line 6,479 of your *Leaves of Grass* file:

```
Placard "Removed" and "To Let" on the rocks of your snowy Parnassus,
```

This is a good time to figure out how to go to a specific line using your text editor, by the way. Being able to do this will come in handy when writing programs that are longer than short snippets. It will also be useful to turn line numbering on so that you can see which lines are displayed. In my editor, Atom, you can control whether line numbers are displayed by going to Edit > Preferences, choosing the Editor tab on the left, and checking or unchecking Show Line Numbers. You happen to be able to go to a specific line number in Atom with Ctrl-G. In Geany, the option to go to a particular line number is under the Search menu and is called Go to Line. It can also be activated with Ctrl-L. Because you have dozens of text editors to choose from, take a look at your menu or your text editor's documentation and see if you can enable line numbering and find a quick way of getting to a particular line. Search online to figure this out if you need to. Be sure to have line numbers turned on and to know how to move to a specific line number before you continue.

Here's what our current regular expression, ".*" (exactly those four characters) will select in this line:

"Removed" and "To Let"

Show yourself that this entire sequence of characters is what is matched. Move your cursor to the beginning of line 6,479 and use the Find option with this expression:

".*"

You should see that entire phrase, with four words, selected.

That is, indeed, an opening double-quotation mark followed by a bunch of arbitrary characters followed by a closing double-quotation mark. We might prefer for our expression to match "Removed" first and then to match "To Let", but regular expressions in general do not work this way. The matching that happens by default is *greedy* so that strings that are as long as possible are matched.

We can fix our quotation-matching problem by writing a more precise expression. Right now we're asking for " followed by *any sequence of characters (except newlines)* followed by ". But in fact, we should restrict what characters can occur inside our quotation. Namely, we shouldn't ever have " (the closing double quotation mark) in there. Fortunately, there's a way with regular expressions to say *every character except this one*. It's done using the caret character ^ which means *not*. Therefore, ^" means *not the closing double-quotation mark*. Now that we've moved beyond the dot, we need to define a character class as we did with the five vowels. So we'll begin our expression with [and end it with]. And we will allow any number of these characters, zero or more, by putting the * at the end:

[^"]*

Just one more thing: right now, this matches any sequence of characters that doesn't have a double quote in it. For this to actually match quotations, we'll want the expression to start with a double-quotation mark and end with a double-quotation mark:

"[^"]*"

Put your cursor at the beginning of line 6,479 once again and search using this pattern. You should see that it matches *only* "Removed", the first quoted text. Click Next (or whatever option you have to find the next match) and you should see that the next string that is matched is "To Let". That is the sought-for result. Now, click Next again. You should find that your expression matches:

```
"Is it this pile of brick and

    mortar, these dead floors, windows, rails, you call the church?

 Why this is not the church at all--the church is living, ever living

    souls."
```

This quotation begins at character 42 and runs over four lines. Remember that our earlier pattern was searching only within individual lines. But by changing . to [^"], we changed from *any character except the newline* to *any character except the double-quotation mark*. That means our new expression will allow newlines to appear between the double quotation marks, and what we find can run over several lines. So, in fixing one problem, we've actually also achieved a goal that we had set aside.

There is one more wrinkle we need to consider. Some of the quotations in *Pride and Prejudice* are more than one paragraph long, and start with one or more paragraphs, each of which *begin* with a quotation mark but have no closing quotation mark; finally, there's a last paragraph to these quotations that has quotation marks at the beginning and the end. (Two of the quotations in *Leaves of Grass* are like this, too, although there the units are lines rather than paragraphs.) Conceptually, will our expression work to find these multiparagraph quotations? Think about why it would or wouldn't. Settle on your theoretical answer to this question before you proceed so that you can test your understanding of regular expressions so far. After you have come up with your answer, try out the expression in both files and see if you are right: see if it does or does not properly handle multiparagraph quotations of this sort.

You will need to see what happens in this case. Whatever the situation, our system for identifying quotations is a pretty good start, and it didn't even require writing a computer program. The whole process simply resulted in a regular expression seven characters long. If it seemed like it took a lot of effort, think about how much time and effort it would take you to go through all of *Pride and Prejudice* and all of *Leaves of Grass*, counting every quotation! And now you have developed a general method, one that almost finds every sort of quotation and can be used on other books as well. And what was difficult to do the first time will be much easier as you continue to work with texts, regular expressions, and (soon) their use in computer programs.

As yet, we've only counted how many quotations there are in two texts. But because we have developed an expression to *match* all the text that is in quotations, we can do more. We can see how much of the text in *Pride and Prejudice* and how much of the text in *Leaves of Grass* is quoted. Figuring out how much quoted text there is, or what percentage of a text is quoted text, seems like it could be interesting, as it gives us some insight into how much dialogue (in the form of direct discourse) there is. It seems more meaningful than just looking at the number of quotations, too. Some tag such as "he said" could occur at the end of a quotation or in the middle of it, between phrases, and that would probably be a more subtle aspect of the text than how much direct discourse was present.

To figure out the amount of quoted text, we'll write a program and work in Python, having learned a few things about regular expressions from our text editor's Find dialog.

[9.4] Finding the Percentage of Quoted Text

Now that we're about to use Python to work with files, we need to figure out how to write a program that reads in a file and can access the text in it. To see how this is done, we'll create a simple but relevant program to read in a file and print it out.

Let's start with figuring out how to read and output the text in the file `1342-0.txt`. It's not tough. We need to open the file using the `open()` function, read it using the file's `read()` method, close the file, and print the result. Closing the file frees up some computer resources, but the main reason to close the file is that we're done with it at this point. If we were to try doing something with that file later, we'd be making a mistake and would like the system to produce a helpful error for us.

The following code will do this task, as long as you have your text file in the same directory where you are running Jupyter Notebook:

```
source = open('1342-0.txt')

pride = source.read()

source.close()

print(pride[:1000])↵
```

Actually, the first line may not work. If it doesn't, there are three obvious possibilities. One is that you are running Windows and your system does not assume that text files have a UTF-8 character encoding. You will see a "UnicodeDecodeError," produced in trying to execute the second line, if this is the case. If you encounter this, you can fix the problem by explicitly indicating the encoding in the first line, changing it to the following:

```
source = open('1342-0.txt', encoding='utf-8')
```

If this isn't your problem, I can think of two more that may have cropped up. Maybe Jupyter Notebook was not started in the right directory, or perhaps there's a typo in the filename. Make sure the filename you are using is the right one. Then, log out of your notebook and start it again, ensuring your new notebook is in the same directory as the text file.

The result of these three lines should be a slice of the string with the first one thousand characters of the file.

Python has a regular expression module; import it, now, to be able to use it:

```
import re↵
```

Using this module, we can find all occurrences of a pattern. To do this, we use the findall() method of the re module and we provide two arguments: first, the pattern we want to use, and then, the string we wish to search—representing an entire document, in this case. To pack our pattern up for use in Python, we will prepare it as a special kind of string, a *raw* string. I'll explain why this is best to do later on; for now, understand that this simply involves putting the letter *r* in front of an otherwise ordinary string. So to find all the vowels in pride, we should use the pattern r'[aeiou]' in the following way:

```
result = re.findall(r'[aeiou]', pride)↵
```

This finds every vowel and places each one, in the order found, in a list—a very long list, in this case. Check the length of it:

```
len(result)↵
```

Were you startled to find that your Python program finds fewer vowels than your text editor did? I was, initially. Then, I realized that this Python code is case-sensitive; it's only counting the lowercase vowels. There are a few ways to change this; one can, for instance, set a flag to indicate whether the search is being done with or without regard to case. But let's do something quick and easy, as we did when we first searched for uppercase and lowercase vowels in our text editor:

```
result = re.findall(r'[aeiouAEIOU]', pride)↵
```

This just "manually" adds the uppercase vowels as well. Check the length of result again to see that you have the same number of vowels as counted previously.

Now, let's take a look at what some of these vowels are. Type:

```
result↵
```

After a moment you should see a list displayed. You can scroll through it; it has all the vowels found in it—more than two hundred thousand of them. This might be an overabundance of vowels. Recall that using slices, we can see just a section of the list. Let's look at the first twenty:

```
result[:20]↵
```

That's a little more tractable. If you look at the file itself, you'll understand that these aren't the first twenty vowels from *Pride and Prejudice*; they're the first twenty vowels from the brief Project Gutenberg note at the beginning. To get *only* the novel, we would need to either write code that ignores the additional front matter and end matter or somehow clean the files. But you can see that with that step done, we could use regular expressions to accomplish some analysis of the file.

Okay, now let's use `findall()` to locate all the quotations. For our purposes right now, we'll include quotation marks themselves when we figure out how long all the quotations are. If that doesn't seem like the right way to do it, there are several ways to take a different approach, but let's keep things simple for now.

We just need to use the regular expression that we already have developed within this Python method, `findall()`. We can take our expression, "[^"]*", and enclose it with the ' characters that we use to indicate strings in Python:

```
quotes = re.findall(r'"[^"]*"', pride)↵
```

Check on how many quotes there are and view the first ten:

```
len(quotes)↵
```

```
quotes[:10]↵
```

Assuming that all the quote marks are paired correctly, if we are interested in knowing the total amount of text that is quotation, we can simply figure out how long all of the 1,751 quotations are when put together. (You did get 1,751 quotations, yes?) We could iterate through the list and count the length of each, adding it to a running total. Or we could just join everything into a single long string and take its length. Be aware that '' here is the empty string, two single straight quotation marks one right after the other:

```
len(''.join(quotes))↵
```

Not bad. What portion of text is that?

```
len(''.join(quotes)) / len(pride)↵
```

To get that as a percentage, just multiply everything by 100:

```
(len(''.join(quotes)) / len(pride)) * 100↵
```

We can stash this value in a variable:

```
pride_percent = (len(''.join(quotes)) / float(len(pride))) * 100↵
```

Now, try the search on *Leaves of Grass*, adding the encoding parameter if necessary:

```
source = open('1322-0.txt')
```

```
leaves = source.read()
```

What else? Call `source.close()` because we're done with the file. Then, let's just display (or *print*) the first one thousand characters to confirm we loaded *Leaves of Grass*:

```
print(leaves[:1000])↵
```

Okay, now the same process for this text:

```
quotes = re.findall(r'"[^"]*"', leaves)
```

```
leaves_percent = (len(''.join(quotes)) / len(leaves)) * 100↵
```

You can see quite precisely what may have been obvious to you beforehand: that the amount of *Pride and Prejudice* that is quotation is much more than is the amount of quoted text in *Leaves of Grass*. This might lead someone to wonder if novels generally use more quotations than poems. Maybe not: Poe's "The Raven" is, using our method, more than 37 percent quotation. That's a lot of quotation for a poem that has only one person and an imaginary bird in it. There are some poems with even more quotation, along with some novels that completely lack quotations. But perhaps there is a trend. Or maybe female authors use quotations in their writing more often than male authors. Or maybe these things are true for certain historical periods and not others.

This example isn't meant to lead to a dramatic conclusion, to a book full of revelations, or to a digital humanities project. The point is that even simple regular expressions (just a few characters in length) combined with a small amount of programming ability can allow one to inquire in new ways, to think about texts in new ways.

And these sorts of techniques have implications for literary art as well. During the first National Novel Generation Month (a computational spin-off from National Novel Writing Month), Leonard Richardson generated an extraordinary text called *Alice's Adventures in the Whale*, which contains passages such as this:

> Alice was beginning to get very tired of sitting by her sister on the bank, and of having nothing to do: once or twice she had peeped into the book her sister was reading, but it had no pictures or conversations in it, "Can't sell his head?—What sort of a bamboozingly story is this you are telling me?" thought Alice "Do you pretend to say, landlord, that this harpooneer is actually engaged this blessed Saturday night, or rather Sunday morning, in peddling his head around this town?"

> Presently she began again. "Ka-la! Koo-loo!" (she was rather glad there WAS no one listening, this time, as it didn't sound at all the right word) "Stand up, Tashtego!—give it to him!" (and

she tried to curtsey as she spoke—fancy CURTSEYING as you're falling through the air! Do you think you could manage it?) "Stern all!"

"My God! Mr. Chace, what is the matter?" said poor Alice, and her eyes filled with tears again as she went on, "we have been stove by a whale." cried Alice, with a sudden burst of tears, "NARRATIVE OF THE SHIPWRECK OF THE WHALE SHIP ESSEX OF NANTUCKET, WHICH WAS ATTACKED AND FINALLY DESTROYED BY A LARGE SPERM WHALE IN THE PACIFIC OCEAN." (Richardson 2013)

In case the remarkable nature of Richardson's enterprise is not evident: he created this new book by writing a program that replaced all of the dialogue—that is, all the quotations—in *Alice in Wonderland* with quotations of similar length from *Moby-Dick*. Of course, at the core of this project was his program's computational identification of which parts of the text were quotations, the same ability we just developed using regular expressions and a tiny bit of Python code. In chapter 15, "Text III: Advanced Text Processing," you will be invited to do a similar sort of project, using not only what you have learned about regular expressions but also the capabilities of the TextBlob library, which are covered in that chapter.

[9.5] Counting Words with Regular Expressions

The hard part of the chapter is done. All that remains is to apply regular expressions to some other problems that have already been encountered. There are a few additional symbols to introduce, but not many.

Let's return to the problem of counting words. Recall that we found all strings that were separated by whitespace by using the `split()` method. Specifically, if we wanted to split *Pride and Prejudice* into words, and we are continuing the Python session from earlier in which `pride` contains the entire text of the novel, we could just do this:

```
pride_words = pride.split()↵
```

Let's see how many words there are:

```
len(pride_words)↵
```

Very good. But we might decide that this method isn't the best one. This method counts *over-scrupulous* and *three-and-twenty* as single words. Furthermore, if there's a punctuation mark (or more than one) at the end of a word—such as a comma, a period, or an exclamation point—that is included with the word. Of course, that doesn't affect the *word count* in its simplest sense, but if we start attempting to determine other statistics (such as average word length), it *will* affect our analysis. So let's shift our definition of a word. Instead of asserting that a word is whatever lies between whitespace, let's say that a word is a sequence of letters. The lowercase letters are [a-z] in regular expression syntax, and both upper- and lowercase letters are [A-Za-z] (or [a-zA-Z] if you like; it doesn't matter if you put the uppercase or lowercase letter range first). One or more of these letters is, for us, a word. Recall that if we want zero or more, we should use *. For *one or more*, there's another special symbol, +. So, try this:

```
pride_words_2 = re.findall(r'[A-Za-z]+', pride)

len(pride_words_2)↵
```

Now we have a few more words. This shouldn't be a surprise because we were expecting that where we had *three-and-twenty* in the previous list, we would have three different words in the new list. If you check out a slice in the middle of the novel (e.g., pride_words_2[4000:4150]), you can see that words are now stripped of their punctuation. Even *Mr.* is reduced to *Mr*—and the possessive *s* is split off as a separate word too.

In addition to the programming experience that you gain, there's an important point here. One method of counting words is not inherently better than another method. It depends upon how you have defined a word and whether that definition is a sensible one for the type of research or art you are undertaking. It also depends on whether your definition of a word is consistent with previous scholarship. In the analysis we have done here, we are taking a particular perspective and are being strictly *lexical*. We are using the properties of characters to figure out where the words are. Is *boathouse* one word, and should it be even if we spell it *boat house*? From a cognitive standpoint, and from the standpoint of orality, it might make sense to count both spellings as being one word. But we can't easily implement this on the lexical level unless we create a special case for that one phrase or word and for every such phrase or word. Any *lexical* method will count the phrase with a space in it as having two words.

The most important matter here is that if you intend to count words and compare your results to those of another researcher, you had better make sure you are using the same definition of a word. Replicating some of the other researcher's results is a good way to confirm this.

This discussion of counting words will be continued in chapter 15, "Text III: Advanced Text Processing."

[9.6] Verifying Palindromes—This Time, with Feeling

Now, armed with knowledge of regular expressions, we're ready to fix our palindrome validation process so that it considers only letters and numerals, not punctuation marks and whitespace. What we will do is prepare the string to be checked so that all the punctuation marks and whitespace have been removed from it.

To get this working, we'd basically like to remove everything except letters and numerals. Letters, numerals, and the underscore character _ are referred to in regular expression terms as *word characters*. Everything else is a *nonword character*. And we have ways of indicating both: \w is a word character, whereas \W is a nonword character.

So one way of describing what we'd like to do is as follows: replace all nonword characters with nothing.

We have a way in Python of replacing all occurrences of some pattern with a string. It's the sub() method of re. And we know that what we'd like to replace is one or more occurrences (+ is the correct quantifier to use, remember) of nonword characters (\W) with nothing. *Nothing*, in this case, is written out as two straight single quotes ' ' because it is the empty string:

```
test = 'Here is a nice string. La la la.'

re.sub(r'\W+', '', test)↵
```

That looks pretty good. We can bundle it up as a function called prep():

```
def prep(text):

    return re.sub(r'\W+', '', text)↵
```

Now, paste your three palindrome verification functions back in and try them out in the following way to first process the text and then check it:

```
palindrome = 'Star comedy by Democrats.'

pal(prep(palindrome))↵

otherwise = "This is almost a palindrome . . . er, actually it's not."

pal(prep(otherwise))↵
```

In the second case, I used double quotes to surround the string because it has an apostrophe in it. After you try these, you should also check to see what prep(palindrome) and prep(otherwise) return.

This chapter only provides an introduction to regular expressions, meant to be good enough so that learners can continue exploring on their own. There are plenty of good resources online that describe all the quantifiers and special characters (along the lines of \w and \W) that regular expressions have to offer. You can also *capture* parts of what is matched within a regular expression and then do something with those captured sections. There are *anchors* to indicate the beginning and end of lines that can be used as well. In addition to defining your own character classes such as [aeiou], there are standard character classes that can be used. Lots of fun remains; this chapter is just meant to persuade those who want to deal with text, whether from an artistic or a humanistic perspective, that regular expressions are powerful and that using them is within reach of beginning programmers.

[Exercise 9-1] Words Exclaimed

Using the two books we've worked with in this chapter, write (very short) bits of Python code that will locate all *words exclaimed*—that is, every word that is followed by an exclamation mark. Of course, the full exclamation is usually a sentence rather than a single word, but begin by figuring out how to identify individual words that are immediately followed by exclamation marks. After you have lists of such words for both *Pride and Prejudice* and *Leaves of Grass*, figure out what the average length of these words exclaimed is for both books. Finally, bundle up your code into a function that determines the average length of these words exclaimed for any text when that text is passed into the function as a string.

[Exercise 9-2] Double-Barreled Words

Whether we consider them one word or two, we find sequences of characters such as *great-grandsons* in both *Leaves of Grass* and *Price and Prejudice*. For the purposes of this exercise, let's call these double-barreled words and determine that we are only interested in words with one hyphen in them; we won't continue to triple-barreled or higher-barreled words. Using the two books we've worked with in this chapter, write (very short) bits of Python code that use regular expressions to find all double-barreled words. Recall that when you use the hyphen in a regular expression, as in [a-z], it has a special meaning. So if you want to actually match a hyphen character, you need to *escape* it by placing \ before it, like so: [aeiou\-]. That means you are indicating the hyphen itself, not using the hyphen as part of the syntax of your regular expression. The expression [aeiou\-] will match every (full) vowel and will also match the hyphen. After you have your lists of double-barreled words, figure out the average length of these for both *Pride and Prejudice* and *Leaves of Grass*. Finally, bundle up your code into a function that determines the average length of these double-barreled words for any text whenever that text is passed into the function as a string.

[Exercise 9-3] Matching within Text

Grab a text file online (or from your system) that has several occurrences of standardized texts—specifically, dates. You can find a page full of forum posts, for instance.

Often, older ones are better for this purpose because many modern-day forums tell you how long ago something was posted. You can also use your own directory listings, including the -1 option (ls -1) on GNU/Linux or Mac OS X. Write a regular expression that will match these dates but not other parts of the text. Try out your regular expression in a text editor to ensure that it works. The right answer will depend on the date format used in your particular data.

[Free Project 9-1]   Phrase Finding
Do this project at least four (4) times. Having done the previous three exercises, develop a phrase finder of your own that locates a large number of phrases in a book-sized document: hundreds, probably, but at least dozens. I don't mean *phrase* in the most general sense here. "Every phrase used to describe an animal" will be too challenging to detect; "individual words that come before a comma" will probably be too limiting to be interesting. You should figure out what sorts of phrases are compelling to locate and list from your own standpoint. Among those, choose one sort that can tractably be found using regular expressions. Run your phrase finder on several book-length documents, including some others besides pride and leaves. You can find phrases that give you some humanistic insight into texts, if you like, or you can find ones that are pleasing to read, silently or aloud. If you want to stretch into new regular expression territory, look ahead to 9.7, Elements of Regular Expressions, for an incomplete reference.

[Free Project 9-2]   A Poetry versus Prose Shoot-Out
Try several approaches as you work on this project, settling on the one you find most interesting. At this point, you should still have the variables pride and leaves defined as the entire text of the novel and the book of poems. The free project is to analyze and compare these texts in a way that you see fit, using your existing knowledge of regular expressions and expanding that knowledge if you like. You can try to lexically determine the length of sentences (all those references to "Mr." Bennet may make that difficult), for instance. Or you can look for certain words, or sets of words, and compare what percentage of the text they take up in each case. Some of the simplest and most commonly overlooked words (articles or pronouns, for instance) may actually be very interesting to consider; avoid manually assembling large sets of nouns, verbs, or adjectives. Again, you don't need to use techniques that haven't been discussed already, but if you'd like to, check out the very next section.

[9.7]   Elements of Regular Expressions

[9.7.1]

Having introduced regular expressions and a few of the first ways you may wish to make use of them, I will now try to break down the main elements you can include in a pattern (also known as a regular expression) a bit more systematically. Even in this section, I am not aiming to supply an exhaustive list or a reference. This isn't even a thorough introduction. Instead, I'm trying to clarify some of the terminology and give a few more examples of elements in each category. This will, I hope, make it easier for you to look up more about regular expressions online or offline and make better use of them as the need arises.

[9.7.2]   Literals

A *literal* in a regular expression is the most straightforward element: It means, "I literally want to match this particular character or string." For instance, when you searched for all occurrences of a in *Pride and Prejudice*, the a you typed in and searched for was a literal. Similarly, searching for the sequence of characters dance involves using that literal string as your pattern.

When you're looking for a letter, it's easy to specify the literal, but some special characters require a special presentation when they occur in a pattern. The way to express a literal newline is \n, for instance. Because the backslash character, \, has a special purpose in regular expressions—it is used to indicate things like the newline, word boundaries, and so on—it needs to be expressed as \\ when you are searching for a literal backslash. There are other special characters, or *metacharacters*, that are used in patterns. These include [and], which are used to define character classes. They can be expressed literally as \[and \] in a pattern. It's similar for other metacharacters, including the . that stands for "any character."

The reason I asked you to use raw strings such as r'\W+', rather than '\W+', is that the backslash character has a special purpose *both* in regular expressions and in the standard (nonraw) sort of Python string. Either one of the previous strings, raw or nonraw, will work in the previous case, but if you didn't use raw strings and you wanted to match a literal backslash, you'd have to write it as \\\\. What a mess! It's better to just use raw strings when specifying patterns.

If you are only intending to search for literals, you don't need to use regular expressions at all, whether in your text editor or in Python. The power of regular expressions comes from being able to combine literals with other elements.

[9.7.3] Character Classes, Special Sequences

As discussed, square brackets can be used to specify a set of characters, or ranges of characters, that will all match. Thus [0-9] will match any numeral, any decimal digit.

As it happens, there is an abbreviation, a special sequence, for this character range, as with several other ones that are often used. For the digits, it is \d. You don't *need* to remember that; it's just a convenience for those who want to match digits very often. Other useful special sequences include \s to indicate whitespace (*s* is for *space*) and \w to indicate *word characters*—that is, alphanumeric characters. If you want all nonspace characters, use \S. Similarly, \D matches nondigits and \W matches nonword characters. These can be written out in extensive form, as character classes with square brackets around them, but as these classes get more elaborate it becomes more useful to know these short ways of indicating them. There are several other special sequences, including, for instance, \b to match any boundary (hence the *b*) between word characters and nonword characters. As you might expect, \B matches anything that isn't a word boundary.

If you want to match several specific letters, for instance, both *a* and *p*, you can place them directly in brackets: [ap]. We have done this with the five full vowels already. Finally, you can put character classes together with literals to make a pattern. Thus [5-9][ap]m will match *5am 6am 7am 8am* and *9am* as well as *5pm 6pm 7pm 8pm* and *9pm*.

Character classes can have negated characters or character ranges in them: [^0] means *everything but 0* and [^A-F] means *everything but the capital letters A through F*. And, of course, the dot or period by itself stands for *everything but a newline*.

[9.7.4] Quantifiers and Repetition

We have used two common quantifiers, * meaning "zero or more of these" and + meaning "one or more of these." There are two more quantifiers. You can use ? to mean "zero or one of these"; in other words, the part of the pattern to which this quantifier applies is optional. Thus travell?er will match either spelling of this word, *traveler* or *traveller*.

The most complicated of the quantifiers uses curly braces and specifies that the pattern must occur some number of times but no more than some other number of times. Using this quantifier, you can specify the pattern Kha{3,7}n! which will match "Khaaaan!" as well as "Khaaaaaaan!"—and will also match the name of Captain Kirk's nemesis if it has four, five, or six consecutive occurrences of *a*. But it won't match "Khan!", "Khaan!", or any version of the name with more than seven occurrences of *a* in it. In simple cases, you can "manually" build a pattern using the other quantifiers.

For instance, the pattern `Khaaaa?a?a?a?n!` works exactly the same way as that previous pattern. It's messier, but if you want to write something like that to start—and you can get your pattern working—you can then refactor it, as with any other code, and end up with a tidier regular expression later in the process. A final note about this complicated quantifier is you can leave off either of the two numbers. The pattern `Kha{3,}n!` covers all exclamations of Khan's name with three or more occurrences of *a* included, no matter how far the name is extended. If you use `Kha{,7}n!`, the pattern will match from "Khn!" up to "Khaaaaaaan!" because zero to seven occurrences of *a* will be fine.

[9.7.5] Grouping Parts of Patterns

Although we aren't diving into this in this book, it's possible to group parts of a pattern together using parenthesis and this can be very useful to do. There are lots of reasons to group parts of a pattern when you're using regular expressions within a program, but I will focus on what's probably the simplest use of grouping. It allows you to apply a quantifier to whatever part of a pattern you prefer—not particularly a single literal, a single special sequence, or a single character class.

To demonstrate this, I'll provide a pattern that will match any sequence of five or more three-letter words in a text. This is a pattern of specific professional interest to me, as I've been writing a poetry book called *All the Way for the Win* which consists entirely of three letter words. Here is the pattern:

`\b(\w\w\w\W+){7,}`

I've specified that whatever we match needs to start at a word boundary (the `\b`) and then have five or more occurrences of the next part of the pattern, which I've grouped using parenthesis. This group has three word characters in it followed by one or more nonword characters, which will include spaces but also punctuation. I then use the fancy curly bracket quantifier to specify that I am looking for seven or more of what's in the parentheses, occurring in a sequence.

I could improve this pattern in a few ways to make it more general. Right now, it will match "words" like *007* that I don't happen to be interested in because they aren't three-letter words; in fact, that one doesn't have any letters at all in it. But that's a first example of how to use grouping and why it might be of use to you. I was quite taken aback when I used this pattern to search through the Enron corpus of corporate emails, finding that someone had excitedly written the phrase "YOU ARE THE MAN FOR THE JOB" (just like that, in all caps) to extend an employment offer.

Groups can be used for many other purposes, not just finding sentences like this one. For instance, you can specify several and, in your Python program, access what is matched in each group, manipulating the data separately. Do start applying those parentheses when you feel ready to try it out. In many cases, though, you can effectively use several simpler regular expressions in different parts of your program. So don't let the complexity of grouping keep you from exploring what you can do with regular expressions.

[9.7.6]   More on Regular Expressions in Python

Hopefully this chapter, and this section of it, provide a running start. So far, I have not mentioned a few of some of the more often-used aspects of regular expressions. There are additional metacharacters. The ^ allows a match only at the beginning of a line and the $ only at the end of a line. There is also an "or" operator | such that the pattern A|B will match A as well as B—that is, A *or* B are both fine. These and other aspects of regular expressions are all very useful, but there are always limits to what can be explained alongside the rest of the artistic and humanistic computing in this book. For another explanation of how regular expressions work in Python—one that is longer and more comprehensive—I suggest the helpful how-to in the Python documentation for the re library, which should be easier to digest with this chapter as a foundation:

docs.python.org/3.8/howto/regex.html

[9.8]   Essential Concepts

[Concept 9-1]   Explore Regular Expressions in an Editor

You should understand the basic, practical way to develop regular expressions (by trying them out in a text editor) and then understand how to include the ones you have developed in a program.

[Concept 9-2]   Patterns Go Far beyond Literals

Regular expressions open up a wide range of possibilities that go beyond a simple *search string*—a pattern that is a literal string. At the same time, they are a simpler formalism than computer programming in general. Once you have understood the way regular expressions extend the traditional idea of finding a plain old string, and once you are comfortable experimenting with regular expressions, you can develop your skills and strengths further as you explore.

[10]   Image I: Pixel by Pixel

[10.1]

In this chapter we'll look at very simple ways of modifying and analyzing image files in a standard, widely used format. This chapter and the next deal with low-level image manipulation, showing how it can be scaled up to work on large numbers of files (using iteration). The manipulations covered are the same as some of the ones implemented in Photoshop and the Glimpse editor (a free software program for photo editing). In chapter 11, "Image II: Pixels and Neighbors," there is a further opportunity for the analysis of collections of images. In chapter 14, "Image III: Visual Design and Inter-activity," different techniques for drawing lines and shapes, and for producing simple animations, are covered. The work in chapter 14 is done using Processing, an ideal language for computational visual design. Processing is introduced in chapter 12, "Statistics, Probability, and Visualization."

Images can be represented in different ways, but the ones we'll consider in this chapter are represented as grids or rectangles of pixels. This *bitmap* representation is a very common one for images. While there are also vector-based representations and other ways of representing images, everything that is displayed on a modern-day computer screen is represented in this bitmapped way, at least at the final stage of display and often earlier.

The advantage of focusing on low-level, Photoshop-like manipulations is that the programming needed to accomplish these is very much like that needed to analyze certain important properties of images. For instance, we will write a short program to redden images—to add red to every pixel. We will also write a short function to determine the redness of images so one can be compared to another. This means we can iterate through large numbers of images and find the reddest one. This is a technique that can be built upon to do other types of meaningful image analysis.

Another advantage is that this work builds on our existing experience with numbers (integers and floating-point numbers) and with strings and lists. A number is truly a single data point, and just as a point, geometrically, has no length or width, a number by itself also has no dimension. Lists and strings are sequences and are one-dimensional, like lines. We can speak of what is to the right or left of a certain element, but if we do that, it doesn't make sense to speak of what is above or below it. As we encounter images, the dimension increases again. An image, like a rectangle, has both width and height, and it *does* make sense to ask what is to the left of, to the right of, above, and below a particular pixel. One of the ways that programming and abstraction can help us generalize involves generalizing to more dimensions.

[10.2]   A New Data Type: Tuples

To prepare for working with images, we'll introduce a new data type that is used to define the size of an image, the position of a pixel, and the four-part (red, blue, green, alpha) color of a particular pixel.

Haven't heard of the "color" alpha? The alpha value is used to control the *opacity* of a pixel. At the minimum value, a pixel is completely transparent—so it doesn't matter what the other three values are. At the maximum value, nothing will show through. This alpha value isn't of course really a color, but it's treated as a color value.

Tuple is the generalized name for pairs, triples, quadruples, and so on. Instead of just providing a number such as 42, a tuple allows you to pair together 42 and 17 in the form: (42, 17). So, for instance, if you want to specify a coordinate in two-dimensional space, it would be quite a good idea to use a tuple with elements—a pair, such as (42, 17). If you needed to specify a point in three-dimensional space, a triple or tuple with three elements would be great to use—for instance, (7, 6, 14). Try this out in your Jupyter Notebook:

```
twod = (42, 17)↵

threed = (7, 6, 14)↵
```

Now, type twod to see the value you've given to that variable, and, similarly, check out the other variable's value.

You may notice that tuples seem very similar to a data type we have already discussed, the list. In fact, any list can be converted to a tuple and vice versa. But a tuple does differ from a list in certain ways. It has a fixed number of elements and can't grow or shrink as a list can. The values of the elements with the tuple also cannot be changed. Thus, it is *immutable*. I'll explain some more about why this type is useful in a bit, but these are the basic differences.

You can access individual elements of a tuple, just as you can with lists. Remember that we used [-1] at the end to get the last element of a list? Try typing twod[-1] and threed[-1], and make a guess about what will happen before you press Shift-Enter. You can also try twod[0] and threed[0] to get the first element, also known as *element number 0*—the element at offset zero.

So, a tuple is a list-like data type with a fixed number of elements. The other thing I mentioned is that while you can *read* or access individual elements, you can't change those elements by assigning new values to them. You need to build a new tuple if you want one element to have a different value. Try this:

```
threed[0] = 8↵
```

The message you get is quite informative. Try this instead:

```
alist = [1, 2, 3]↵
alist[0] = 17↵
```

You'll see that there is no error. To see what this assignment does, type alist, take a guess about what will result, and press Shift-Enter.

Note that the immutable nature of tuples doesn't prevent us from producing many tuples as we loop or iterate. To see that it's possible to produce numerous tuples by iterating, try these two lines in the interpreter:

```
for i in range(5):
    print((i, 2, 3))↵
```

The outer pair of parentheses belongs to the print() function, while the inner pair shows that what is being printed is a tuple—in this case, specifically, a triple.

Why are these two very similar data types, list and tuple, available in Python? To some extent, it's because it's possible to be more computationally efficient with an immutable data type (such as a tuple) than with one that can be changed (such as a list). But there's another important reason, too, which has to do with people rather than with computers.

Sometimes it just makes sense to define an immutable sequence of values. If we know of a pixel's coordinate within an image, say, (12, 502), we can't remove one of the numbers from it—to get (502)—or add a number to it—to get (12, 502, 27)—and have that coordinate remain meaningful in the way the original pair was meaningful. We probably shouldn't modify just its horizontal or just its vertical position, either; in certain special cases, that makes sense, but in general it's most reasonable to consider that the pair of coordinates is "a thing" and to provide the entire new coordinate. A tuple can do more than just enhance our computer's efficiency; it can help us think about a problem by making certain things, which are seldom sensible, difficult and by making other things, which are often sensible, easy.

This is a major justification for putting type systems in place. They provide a light-weight means of checking programs during their development, providing help to programmers. Although they don't do everything necessary to ensure that a program is correct, they can help. For instance, if you define a Celsius temperature that is 'hello' and then try to convert this to Fahrenheit, you will get an error. See for yourself by quickly typing in a definition of the conversion function. You don't even need to include your custom error:

```
def to_f(c):

    return ((9/5) * c) + 32↵
```

Now, try the conversion:

```
to_f('hello')↵
```

The error message isn't extremely informative in this case, but those who know something about types can see that the problem has something to do with strings and integers. And, essentially, it does: we tried to ask for the Fahrenheit value of 'hello', a string. Python helped us out by producing an error. This was a helpful response because what we were asking was actually nonsensical. Python doesn't know anything about the word *hello* specifically. It was the conflict between different types (strings and integers) that allowed this helpful error message to be produced.

We now know about three different types of sequences in Python: lists, strings, and tuples. We know that double() works if it is given either a list or a string as an argument. Go back to the code for this function (retrieving it from your earlier Jupyter Notebook session and pasting it into your new session) and check to see if double() will work on tuples.

[10.3] Generating Very Simple Images

To help reveal the link between generation and analysis, we'll create some images from scratch. Initially, these will be images that are entirely of one color. We'll use the Python Imaging Library (PIL). This really isn't what PIL was made for; it's intended to allow the processing of images that were created in a drawing program, were produced by a camera, or came from some other source. Still, to get a feel for how it works and how computers work with images in general, we'll use it to program a very simple image generator. It's simple enough that we can do it in Jupyter Notebook.

Although PIL is not part of "core" Python, when you install Anaconda, it is included; no extra installation steps are needed. The first step in writing some code is importing the class we need, Image, from the Python Imaging Library, PIL:

```
from PIL import Image↵
```

There's no outcome, except that we now have access to the Image module. Next, we create our first image, which will be labeled ourimage:

```
ourimage = Image.new('RGBA', (100, 100), 'black')↵
```

This is actually all it takes to create an image. Because we haven't yet saved the image, it exists only as a variable. We can't find it in a folder somewhere and open it up in an image viewer—not yet. But we can see the value of our variable in Jupyter Notebook. Be sure to guess what this will look like before you press Shift-Enter!

```
ourimage↵
```

Here is an equivalent way to write that single line of code that provides labels (variable names) for each of the three arguments to Image.new(). Try typing it in, so you get a bit of practice meaningfully labeling values with variable names:

```
mode = 'RGBA'
size = (100, 100)
color = 'black'
ourimage = Image.new(mode, size, color)↵
```

Please do satisfy yourself that it is effectively the same as the earlier line. Labeling each of the arguments, and then using those labels in Image.new(), makes this version clearer. If you are writing code to share with others or for your own use over time, something like the latter version will be preferable. To poke around, it's perfectly fine to use the shorter version, but even if you are the only person who will read the code, it could help you to write in the more extensive form. After typing out this longer form, view the image to make sure it looks the same as it did before:

```
ourimage↵
```

Now, let's save the result and make sure that our line of code worked:

```
ourimage.save('allblack.png')
```

You should be able to open `allblack.png` (you'll need to find it in whatever directory you are using with Jupyter Notebook) and see the same small black square, 100 pixels on a side. You can use your standard GUI (i.e., you can just click on the icon) to open this up in whatever image-viewing program you usually use.

Here's what is happening in `ourimage = Image.new('RGBA', (100, 100), 'black')` and in the extensive version of the same code. The variable `ourimage` is being declared and assigned a particular value. As with any variable name, what we call this image is up to us; it could have been `fred`. It was just meant to be a somewhat meaningful label. The value assigned to `ourimage` is an Image object, one that is created using the `new()` method of the `Image` module. The `new()` method is being given three arguments here.

The first argument is a string indicating the color mode, which is why we used the variable name `mode` in the extensive form of the code. RGBA (red, green, blue, alpha) is the main color mode we'll use in this book. It's a common and useful one.

The next argument is one of those sequences—the type that was just introduced, a tuple—and has two elements indicating width and height. In the extensive form of the code, we labeled this pair with the name `size`. In our case, the width and height are the same, so the image created is a square.

Finally, the last argument indicates what color the whole image should be. We could use a more cryptic (but in certain ways more versatile) representation of color, and we will in a moment. It happens, though, that PIL understands the string `'black'` to mean the same thing as the hexadecimal triplet with minimum values for red, blue, and green—as little red, blue, and green as possible—and with full opacity, so that nothing shows through it. So though we can't ask for the Fahrenheit conversion of `'hello'`, we can ask for the color `'black'` because this method's interface is written to allow this. Sensibly, we used the variable name `color` to label this when we wrote the code in extensive form.

By now, it's clear that `(100, 100)` means a 100 × 100 square, but which of these numbers indicates width and which indicates height?

You could search the Web to find out, but that would actually take longer than exploring Python and PIL yourself and determining the answer through experimentation. Just go back to the line beginning `ourimage =` and edit it. Change one of the two `(100, 100)` values to something other than 100. For instance, change one of them to 15 or 200. (Changing it to 101 wouldn't be a very good idea because it would be really difficult to see the change that you made.) Press Shift-Enter on that line, and then Shift-Enter on the next line—`ourimage`—to view the newly created image in Jupyter Notebook. You can tell by looking at the change you made which number corresponds to the horizontal or *x* dimension and which to the vertical or *y* dimension. If you forget at any point, create a (200,100) rectangle and see whether it's wider than it is tall or taller than it is wide.

This is a good point to mention some other ways to "ask Python itself" about how it works. To take one example, if you want to know how the len() function works, you can have Python offer help by using the built-in help() function:

```
help(len)↵
```

When you ask for help, you aren't actually calling or invoking the function. You're treating the function as something to be inquired about. For this reason, you need to leave off the parentheses that usually go with len(). What results is brief documentation of the function—and you already know how it works, of course. There is much more elaborate documentation for more elaborate components of Python. For instance, you can ask for help with the entire Image module:

```
help(Image)↵
```

That is probably way too much help to be helpful! Instead, try asking for help with just the Image.new() method:

```
help(Image.new)↵
```

There's some more focused help. The language is a bit technical because it is meant to be precise, but you can see there is built-in documentation. One thing it explains is that the third parameter is optional, and black is the default color for an image.

It happens to be extremely easy to add your own help information when you write a function. Just add a string (officially known as a *docstring*) immediately after the first line of the function definition and before any other code. Try it out with the detective tax function from 6.2, A Function Abstracts a Bunch of Code:

```
def tax(subtotal):
    "Return the amount of detective tax due on the given subtotal."
    return subtotal * 0.08875↵
```

Now see what happens when you ask for help regarding your own function:

```
help(tax)↵
```

Including a docstring can be extremely helpful. You could figure out what the function is doing by going through the (one line of) code, but there is nothing else in the tax() function that would explain why it was written—that this function computes the tax on detective services. This is why comments, generally, are useful. In Python, the most important comment-like text you should write is the docstring that briefly explains the operation and purpose of each function you develop.

You also may want to know, for a specific Python object, what methods and attributes that object has. For instance, what can we do with a list? To find this out, we can ask for the *directory* of a list using the `dir()` function. Any list will do; let's use the empty list:

```
dir([])↵
```

As you can see, this generates a list itself. It's a list of all the attributes and methods that apply to lists. For instance, you can see `count()` in this list, albeit without its parentheses. It's nice to know about it, but this list by itself doesn't tell us what `count()` does. However, we can ask for help to see the docstring of `count()`:

```
help([].count)↵
```

Is that clear enough? Can you try using the `count()` method with the list `[1,5,2,4,1,3,1]` to see how many occurrences of 1 there are in the list? Give it a try in Jupyter Notebook.

Checking docstrings and using `dir()` is not a substitute for a tutorial, or for convenient and more extensive reference documentation, which is easily available online, but it does show you that as long as you have a Python interpreter such as Jupyter Notebook, you have the ability to reflect upon and get some help on the Python programming language itself.

To return to the task at hand, let's look at the alternative, more generally useful way of indicating the color that was mentioned. Try the first line again, the short version of the assignment, replacing the string `'black'` with a quadruple, `(0, 0, 0, 255)`. This indicates the same color (black) in a different way (the minimum amount of red, blue, and green—all zeros—followed by the maximum opacity, 255). For simplicity's sake, when we generate images, they will all be completely opaque; the last of the four color values will always be 255.

Confirm that your code works by viewing the image in Jupyter Notebook. It doesn't matter what the dimensions of the image are, so long as it's large enough for you to see it.

Next, try changing the color from `(0, 0, 0, 255)` to `(127, 127, 127, 255)` to produce a medium gray instead of black. Assign the image you have generated to the variable `allgray`. If you use a different dialect of English or would like to affect using a different dialect of English, you can call this variable `allgrey`. By the way, PIL will helpfully understand both `'gray'` and `'grey'` to mean the color `(127, 127, 127, 255)`.

Now, go back to the line beginning `ourimage` = and make the image redder by increasing the first number in this tuple. Specifically, use (`200, 127, 127, 255`) as the color. Assign this image to the variable `allrose`. Already, with this simple step, and even though we are creating a completely uniform image, we are making use of a color that, unlike `'white'` and `'gray'`, is not represented by PIL as a string. Using tuples, it is easy to access any of the 256 × 256 × 256 colors (not counting the alpha values) that typical contemporary computers support.

You can change the individual pixels of the image by setting them to different quadruplets. Try this:

```
allblack = Image.new('RGBA', (100, 100), (0, 0, 0, 255))↵

allblack↵
```

Now, remember that when we looked at the help information, it explained that black is the default color for new images? Try to create the same rectangle with no color specified to see if it gives the same result:

```
default = Image.new('RGBA', (100, 100))↵

default↵
```

This is a strange result! Let's use a special means of reading one particular pixel—we'll pick the one in the upper left corner, at (0, 0)—to try to figure out what happened:

```
default.getpixel((0,0))↵
```

Aha! When you see this quadruple, you should be able to figure out that the image *is* actually black. The red, green, and blue values are 0. However, it's also completely transparent because the alpha value is 0 as well. So it's a very odd type of black. If we were using the RGB color mode, we wouldn't have run into this issue.

Let's go back to our manually created version, now that we've discovered this wrinkle. Now we'll set a different color for a central pixel:

```
allblack.putpixel((50, 50), (255, 255, 255, 255))↵
```

Take a look at the result. If the change isn't clear, you can save the image, open it in your standard image viewer, and zoom in on the middle of it:

```
allblack.save('almostall.png')↵
```

You should be able to see (if your display and your eyes are both working very well) that there's a single white pixel in the middle of this image. That pixel at coordinate 50, 50 has been set to have the maximum red, blue, and green values (all 255) and also is completely opaque (255 for the last element, the alpha channel).

Recall that the computer can do things one hundred times about as easily as it can do them once. And recall that you can generate a sequence starting at 0 and ending with 99 using `range(100)`. With this is mind, can you write a program that extends this single-pixel bit of code so that a horizontal line is drawn across the middle of the image? Drawing the line can be done in the interpreter with two lines of code, although you need to use some existing image (such as `allblack`) or create your own, and after you're done drawing the line you need to save the image as a file.

Here's a hint about how to proceed, if you can't think of a first step to take. First, realize that you *could* write a hundred lines of code to accomplish this in a straight-forward but certainly not very elegant way. It would involve writing something very similar to `allblack.putpixel((50, 50), (255, 255, 255, 255))`, but instead of `(50, 50)`, the first line would have a different coordinate, the second line would have yet another coordinate . . . and the last line would have yet another coordinate, too. Without actually writing all one hundred lines (which would be very tedious, even using some sort of copy-and-paste method), think about what those lines would be. Then, write a loop that, using iteration, will do the same thing in *two* lines rather than in one hundred.

You'll want to finish by saving the image as `oneline.png`. The process is the same as that for saving the earlier image file. You'll invoke the method `allblack.save()` (assuming you have called your image `allblack`) and will also need an argument, the filename that you want to use.

It's good to know that a system for modifying and analyzing images can also be used to generate them from scratch. Using this capability provides the opportunity for some low-level pixel setting. In chapter 12, we will begin to work with a programming language that was made for drawing images from scratch: Processing. You'll see how much easier it is to draw lines (and do much more) anywhere on the image that you like, in any orientation. And we'll use these sorts of techniques to create not only images, but also, in chapter 14, animations.

Our use of PIL to draw an image pixel by pixel brings up another important point about programming and computing: what is *possible* isn't always a good idea to imple-ment. In working on projects, questions that nonprogrammers ask of programmers often begin with, "Is it possible to . . . ?" But the answer to this question is seldom very informative. It is *possible* to generate an elaborate computational drawing, such as an image of a fractally branching tree, using Python and PIL. Unless you're setting this as a personal educational challenge for yourself, however, this is almost certainly a bad and perhaps even a boneheaded idea. In Processing, this sort of image can be generated in a few lines—and there are even easily found, well-known examples of how to do this on the Web. PIL is made for automatically doing the sort of image processing that Photoshop does and is good for image analysis, but it is not best for drawing images from scratch according to computational rules.

If you have the chance to collaborate with professional programmers on specification-based projects, I suggest beginning these sorts of questions with something like, "How difficult would it be to . . . ?" In some cases, it might be best to follow up by discussing the question, "Would it be more sensible to use a different programming language, library, or framework?"

[10.4]   Pixel-by-Pixel Image Analysis and Manipulation

For simplicity's sake, even though we created one nice, rose-colored rectangle, we'll begin this section by discussing grayscale images, images in which all the color values have the same amount of red, blue, and green. We'll use images of the same standard format, PNG. Leaving color for later is not an unusual way to proceed. In programming, it's often good to start with simpler versions of a problem. In visual art and design, intensity (dark and light) is usually dealt with before color.

In a PNG image, the intensity of each pixel varies between 0 (no light, a completely black pixel) and 255 (the maximum amount of light, a white pixel). It might be easier to think about if the range went from 0 to 100 and we could simply specify a percentage of light, but the 256 possible values of intensity correspond to the values that can be stored in a single byte, which has eight binary digits (or bits). These bits can each be 1 or 0, on or off, and because there are eight of them in order, there are 2^8 or 256 possibilities. If we start counting from 0, our first value, then 255 is our 256th value, the maximum number.

It isn't essential to know about the architecture of computers to understand how to program them effectively; one of the types of *abstraction* that developers of programming languages have striven to implement is the ability to think about computation and problem-solving independent of hardware. It can be good to know that when 255 and similar values crop up, however, they are not being placed capriciously or maliciously by the creators of different systems. They actually relate to underlying aspects of computer systems. You will get to learn more about this in chapter 16, "Sound, Bytes, and Bits."

Some programming environments do allow for percentages or similar values to be specified instead of numbers in the range from 0 to 255, and this can be a help to new programmers. John Maeda's Design By Numbers (DBN), a predecessor of Processing, works this way. Even in this age of high-level languages and abstraction, however, there are very good reasons to use the computer's native representation. If one can only specify an integer value between 0 and 100, any mapping that is used will be uneven and it will not be possible to indicate many native color values (the other 155 of them). If one is allowed to specify a floating-point number between 0 and 100, a conversion between floating-point and integer numbers will have to take place—every time a pixel is updated. Also, it will probably be necessary to store and work with a grid of 0 to 100 color values apart from the underlying native values. As more intensive computer graphics processes are added to this basic method of updating pixels, performance can suffer. Plus, the added complexity of a system like this can provide a challenge in the development of such a system in the first place and can make maintenance of such a system difficult.

The image manipulations discussed here are ones that affect the entire image. In other words, we're no longer concerned with drawing a single pixel in the middle of an image or a one-pixel line across it. We would like to deal with global properties of an image.

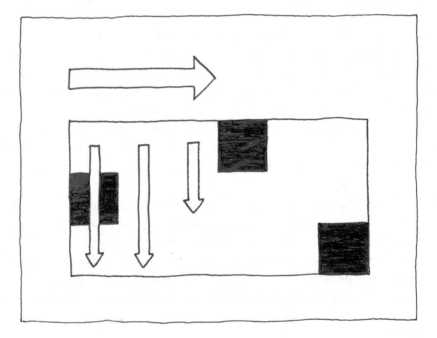

[Figure 10-1]
Nested iteration is an easy way to visit each pixel in an image. Here, the outer loop proceeds horizontally; in this case it has gone through column 0 and 1 and is currently on column 2. At each step of the outer loop, an inner loop proceeds through each pixel in the column. Inner loops already visited all three pixels in column 0, all three in column 1, and the first 2 in column 2.

To do so, it won't be enough to change one pixel or (in a 100 × 100 image) only one hundred pixels. We'll need to visit every pixel. We can do that easily with a *nested loop*. The inner loop will go through every pixel in a column. The outer loop will go through every column in the image. It's just iteration, of the sort we have already covered, but now the iteration goes through two dimensions instead of just one.

To get such a loop working, let's simply go through every pixel of an image and set each pixel to gray—the color value (127, 127, 127, 255). We can imagine the same sort of processes happening if the world ends up being destroyed by a nanotechnology accident. This only takes a few lines, so let's try it in the interpreter. There is no need to type these next three lines in again; just review them to recall how we created an image, set the middle pixel to white, and saved the image:

```
allblack = Image.new('RGBA', (100, 100), (0, 0, 0, 255))↵
allblack.putpixel((50, 50), (255, 255, 255, 255))↵
allblack.save('almostall.png')↵
```

To go through the entire image, it really makes sense to use two loops. If you drew a line across the image earlier, you've already written half the code that is needed. This is all the needed code, written to work with an image that is 150 pixels wide and 100 pixels tall:

```
rectangle = Image.new('RGBA', (150, 100), (0, 0, 0, 255))↵
rectangle↵
for x in range(150):
    for y in range(100):
        rectangle.putpixel((x, y), (127, 127, 127, 255))↵
rectangle↵
```

If you just opened a blank Jupyter Notebook for your work on this section, you'll find that Image is not defined after you input the first line. Remember, it's necessary to enter and run from PIL import Image at the beginning of each session. If you kept your previous session open, Image will already be imported from the PIL module.

There are five lines in the code to create this image and set it, pixel by pixel, to gray. The first and last are just like ones we've already seen, but with different names for the Image object and the file and with a different image size. The second-to-last line is just like the earlier line that sets the color value of a particular pixel, except that instead of setting pixel (50, 50), it sets (x, y), and it sets the color to gray, (127, 127, 127). So the real questions are what the variables *x* and *y* stand for (what values they have) and what the two for loops are doing. These are actually the same question.

The loop specified by `for x in range(150):` goes from 0 to 149, assigning each value to *x* in turn. So *x* is first 0, then 1, then 2 . . . and eventually 149, just before that loop ends.

Within that loop, `for y in range(100):` loops from 0 to 99, assigning each value to *y* in turn. So *y* is initially 0, then 1, then 2 . . . and finally 99. And *y* is assigned *all* of these 100 values for each value of *x*.

That means that `putpixel()` is first invoked with *x* having the value 0 and *y* having the value 0, which we can call (0, 0). The next pair is with *x* 0 and *y* 1, or (0, 1). Then, (0, 2), (0, 3), and so on up to (0, 99). And that takes care of the first column, the vertical line (going along all values of *y*) on the far left. After this, the value of *x* increases from 0 to 1 and `putpixel()` is called with (1, 0), (1, 1), (1, 2), and so on up to (1, 99).

If this isn't clear, try expanding `range(100)` into a list by using the `list()` function to cast it to this data type. You should imagine what this expansion looks like. Then, type the expression `list(range(100))` into the interpreter and see if you're right. Having done this evaluation, this expansion, you're looking at all the values that y takes on at each step through the loop with `for x in range(150):` at the top.

Of course, it's much easier to create a gray image simply by using `Image.new()` and by giving that method (127, 127, 127, 255) as the last argument! If you know you want a gray image to start with, that's the way to do it. But we're not trying to create a new image that is colored entirely gray; we're trying to learn how to go through an image pixel by pixel.

To show how this method of coloring applies generally, let's try it again with one small change. Instead of coloring every pixel (127, 127, 127, 255), let's color each pixel (x, x, x, 255). Now, the conventionally named x is the variable that tracks the horizontal coordinate of the image, what is typically the x-axis. What this means is that at pixel (0, 0), the value of *x* is 0, and this pixel will be colored (0, 0, 0). At pixel (1, 0) the value of *x* is 1, and this one will be colored (1, 1, 1)—and so on. Pixel (2, 0) has *x* = 2 and will be colored (2, 2, 2), and pixel (100, 0) has *x* = 100 and will be colored (100, 100, 100). The same coloring will be done all up and down the image (along the vertical y-axis) because the value of the variable y doesn't have any effect on how an image is colored.

Think about what you expect the image to look like and then run the following code. After doing so, open `gradient.png`:

```
for x in range(150):

    for y in range(100):

        rectangle.putpixel((x, y), (x, x, x))↵

rectangle.save('gradient.png')↵
```

Does the image contain black? Does it contain white? Can you change the code so that the gradient goes the other way or along the other axis? Please do!

Before continuing, let's see how a similar function can tell us one simple, global property of an image: how red it is. For our purposes, we'll define *redness* as the difference between the total red values and the average of the blue and green values. Here's this analytical function, ready to work on the square images, 100 pixels wide and 100 pixels high:

```
def redness(square):

    all_red = 0

    others = 0

    for x in range(100):

        for y in range(100):

            red, green, blue, alpha = square.getpixel((x, y))

            all_red = all_red + red

            others = others + green + blue

    return all_red - (others / 2)↵
```

After typing this in, try it out on `allrose.png` and `allblack.png`, the square images you created earlier. It should return a positive number if the red values of all the pixels exceed the average green and blue values. In an image that is all black, where red, green, and blue values are equal everywhere, the answer should be zero.

This is a fine analytical function, for what it is. But it isn't general. The image-generating code laid out previously, and this `redness()` function, both are hard-coded to images of a particular size. We'll address that issue in the next section.

[10.5] Generalizing to Images of Any Size

Having gone through the image `rectangle` pixel by pixel, we can generalize the technique for doing this. We'll generalize it in two ways: First, by making it into a function.

Then, we'll determine a way of going through each pixel that works no matter what size the image is. It wouldn't be very useful to have these particular width and height values hard-coded and to only be able to deal with 100 pixel × 100 pixel images, particularly if we expect to process directories full of different images.

It happens that the size of an image is supplied by Image objects, in an attribute; try typing:

```
rectangle.size⏎
```

Just to make sure this works generally, create another image that is larger, and then try asking for its size:

```
bigger = Image.new('RGBA', (500, 600), (0, 0, 0, 255))⏎
```

```
bigger.size⏎
```

Great; that's the size of an image, provided to us in a tuple—specifically, a pair, with the width first and then the height. Notice that we haven't saved this image yet by invoking the save() method, so we haven't made it available via the file system. However, it exists in memory here in the Python interpreter, so we can still inquire into what size it is and do other things with it—including, eventually, saving it. When we check the size attribute of an image, we're asking for the size of an Image object in memory, which might have been loaded from disk or (as in these cases) newly created.

The first element, element number zero, is the width:

```
bigger.size[0]⏎
```

And of course the second element, element number 1, is the height:

```
bigger.size[1]⏎
```

The same of course will work for rectangle as well as bigger; it will work for any Image object. Knowing this, you should be able to create a *general* version of the code that we wrote to visit every pixel and turn each pixel gray. The code we have works for images 150 pixels wide and 100 pixels tall. Let's bundle the code we have for this into a function:

```
def grayout(pngimage):

    for x in range(150):

        for y in range(100):

            pngimage.putpixel((x, y), (127, 127, 127, 255))⏎
```

Try it out in Jupyter Notebook. You'll need to define a new image, called whatever you like, and then pass it to grayout(). Finally, you'll need to save the result. All of that will go like this:

```
test = Image.new('RGBA', (150, 100))↵

grayout(test)↵

test↵
```

Now that the code is conveniently bundled up, we're ready to try generalizing the code so that it works on images of any size. We know our hard-coded width is right there at the end of for x in range(150): and we also know the way to get the width of any image (using the size attribute). So, shouldn't we replace the former with the latter? And shouldn't we do the same with height? See how you can make these substitutions. Create a new image of some other size and test your modified, general function grayout() on that image. You can use the image bigger, if you like, or another image you create. Whatever image you use should have a different width and height as in the initial grayout(); otherwise, it won't make for a very good test case.

[Exercise 10-1] Generalizing redness()

After completing the modification of grayout(), also modify redness() so that it works on images of any size. First, the hard-coded values for width and height should be replaced in the exact same way with the width and height as provided by the size attribute.

There is an additional step for this analytical function, however, now that images of different sizes are going to be compared. The result it returns needs to be divided by the area of the image. Otherwise, a large image the same color as a smaller one will end up seeming more red, according to this function.

Finally, remember to have your code, and the names of variables and parameters, accurately represent what computation is happening and what data is being handled. In the original redness(), there is an argument named square, but in your modified function this image does not have to be a square. Give it a more suitable name.

Create some images in Jupyter Notebook, of different sizes and with different amounts of red in them, to try out your function.

Having written a new grayout() and a new redness(), you have frameworks that will be effective for many different types of image manipulations and for very high-level sorts of image analysis. Your grayout() code traverses every pixel and makes a change—in this particular case, just setting each pixel to gray, but, when modified, doing something more interesting and useful. The code will need to be changed a bit for some manipulations, as you will often wish to get the current pixel's color values rather than simply overwriting them—just as is done in redness(). As is, these two snippets of code represent important ways to process images.

[10.6] Loading an Image

So far, we've created images and then modified them. Let's use our very simple image-modification code, which sets everything to gray, on an image that already exists. To do so, we'll need to load such an image, and of course we'll want to check to see that we did actually load the image.

For image manipulation purposes, obtain a PNG image from the Web: the search term "PNG" will probably be helpful. Look for something with some detail and with different levels of brightness and for an image that isn't horizontally or vertically symmetric. A photograph or a scanned image of a painting or drawing would work; line drawings that have only black and white or images that consist only of words are not best. Rectangles of all one color will be a particularly bad choice.

Use a Jupyter Notebook that was started on your desktop.

To begin, if this is a new session, type from PIL import Image.

Then, put your PNG image on your desktop. Rename it, using example.png as the new name. If you started Jupyter Notebook on the desktop and placed the image there, this will open it:

```
ourimage = Image.open('example.png')↵
```

That should load your image. If it doesn't, the image with this name is not in the same directory that Jupyter Notebook is using. If this expression did not produce an error, view the image:

```
ourimage↵
```

Now, if your generalized version of the grayout() function isn't already part of your active session, type it in or paste it in again and press Shift-Enter. Then, try this:

```
grayout(ourimage)↵
```

This should set every pixel in the image to medium gray. It will fail if the PNG image uses a different color model. For instance, some PNG images are not transparent, and some are grayscale images. Most of the PNG images on the Web are of the RGBA sort, however. If you found a PNG image that uses a different color mode, just go back to your browser and keep searching until you get an image that works. After this runs without an error, enter:

```
ourimage↵
```

This will show you the result. Check to see if this is an entirely gray rectangle of the same dimensions as the original image.

[10.7] Lightening and Darkening an Image

At this point, we have a framework for doing all sorts of image manipulations, no matter what the size of the image, and we know how to load a PNG image for manipulation, no matter what that image is. So, we're ready to try a real image manipulation, of the sort that would be done by a program like Photoshop. We will adjust an image to make it lighter.

Let's load our example image once more. If we didn't, we would just have a gray rectangle in the bucket of the ourimage variable.

```
ourimage = Image.open('example.png')↵
```

Now, let's develop a new function that reads every pixel of an image and, for now, writes the same color values right back to the image. This will leave the image unchanged, of course. To do this, we'll need to *read* a pixel, or *get* its color values. We've previewed this already, so try using the getpixel() method on the image, once again, and pass this method (0, 0) to check the color values of the upper-left pixel:

```
ourimage.getpixel((0, 0))↵
```

Because you have already found a PNG image that worked with grayout(), you will find that the value returned has four elements (red, blue, green, alpha). This is not the only possibility for what getpixel() might return, but we will assume this color model for now. Also, let's keep in mind that even when we start changing the red, blue, and green values, we don't want to manipulate the transparency.

Now, to sketch a function that will process the image:

```
def modify(pngimage):
    for x in _____:
        for y in _____:
            (r, g, b, a) = pngimage.getpixel((x, y))
            pngimage.putpixel((x, y), (r, g, b, a))↵
```

You'll have figured out what to iterate over previously; include that, for both horizontal and vertical dimensions, in the two blanks.

This is a somewhat misnamed function right now because it doesn't modify the image—or, you might say it performs the *null modification*, replacing each pixel with exactly the same pixel as before. The assignment (r, g, b, a) = pngimage. getpixel((x, y)) is a clever way of assigning four variables at once: getpixel() returns a tuple, specifically a quadruple, and this allows us to assign its four elements to the variables r, g, b, and a.

Try this function out to make sure it doesn't change the image and to confirm your understanding of the code. A function that isn't supposed to do anything may not be a very sophisticated test, but a mistake in entering the function could be revealed by trying it. If there is an error, or if this function *does* change the image, then there's a problem that needs to be corrected. You can, and should, also check the file modification date to see that the image is in fact being rewritten, even though it is being rewritten to be the same as it was before.

Next, let's try a simple modification of modify() that will cause it to lighten images a bit. Recall that the minimum color value is 0 and the maximum (brightest) is 255. So if we wanted to make an image lighter or brighter, we would add something to each color value. And we probably wouldn't want to add a small value such as 1 or 2, because it would be hard to notice that change. Let's try adding one-fourth of the range of intensities: 64. All we need to do is add 64 to each color component as we put the new pixel.

What's the easiest way to go about doing that?

You can, if you like, modify the values of r, g, and b by including new lines that add 64 to each. The first one will read r = r + 64 and the other two will be the same except for the use of g and b. But there's an easier way. Try changing the last line to:

```
pngimage.putpixel((x, y), (r + 64, g + 64, b + 64, a))
```

Remember, we don't wish to change the alpha value. If you run the new modify() on an image, and then save that image, you should see that it is considerably brighter. But one thing that you won't see, if you develop modify() as described previously, is an error. And that seems a bit odd.

What if one the pixels you're adjusting is completely white—with color values (255, 255, 255, 255)—to begin with? It certainly can't be brightened in that case. An attempt to lighten or brighten it would mean attempting to set the color value to (255 + 64, 255 + 64, 255 + 64, 255) or (319, 319, 319, 255). That's not right! Pixels can't have that value, any more than they can have the value 'hello' or 3.14. For each pixel, there are three color values, and the color value is at least 0 and at most 255. At a low level, trying to assign a number larger than 255 is like trying to put more than one bucketful of water into a bucket or expecting a single coin to come up three heads at once. There is simply not enough to hold the value—fundamentally, no circuitry with which the larger number can be represented.

Now, while a pixel can't have the value 'hello' and also can't have the value 'white', we can use the string 'white' with putpixel() to set a pixel's color values. putpixel() determines that the string 'white' means (255, 255, 255), and it sets the color values that way.

Similarly, putpixel() also figures out that when given (319, 319, 319), it should set the pixel's color values to (255, 255, 255). If any of the color values in the triple *exceed* 255, putpixel() will just use 255 instead. If any of them are negative, putpixel() will use 0 instead of those values.

This is a form of *saturation arithmetic*. It is not always sensible, but it makes a great deal of sense for making changes in color values. If putpixel() raised an error whenever the value was less than 0 or greater than 255, we would need to cover all of these cases in our own code. After setting r = r + 64, we would have to test r to see if it was greater than 255; if it was, we would need to set it to 255. And then we'd have to do the same thing for g and b. That would be a lot of additional code. And yet when we end up with a value greater than 255 with our r + 64, that doesn't particularly mean that we made a mistake. We just lightened our image into the as-much-red-as-possible range. Here, putpixel() is being "forgiving" in a particular way, but in a way that is unlikely to cause problems and is much more likely to help us. By the way, the same saturation arithmetic applies to the alpha value, although because we aren't changing it, we aren't concerned with this.

To show that we can generalize this function without too much trouble, I will offer a version that works whether or not the original image has transparency—that is, whether we have an RGB or an RGBA color model. The only reason I've bothered using RGBA mode, which adds a bit of clutter, is that PNG images you'll find online are often in this color mode. To be as general as possible, we'd like our function to work regardless of which mode the image uses. Here's one way to do that:

```
def modify(pngimage):

    for x in _____:

        for y in _____:

            (r, g, b) = pngimage.getpixel((x, y))[:3]
```

```
new_color = (r + 64, g + 64, b + 64)

new_color = new_color + pngimage.getpixel((x, y))[3:]

pngimage.putpixel((x, y), new_color)↵
```

The first change is the addition of [:3] in line 4. This indicates that we are taking a *slice* of the pixel's color tuple, obtaining the first three elements. Slices are useful beyond text manipulation, as you can see here. In the next line, new_color is assigned the updated red, green, and blue values that have been lightened. Now, when we end up sending new_color along as an argument, we would like it to have three elements (if the original pixel had three) and four otherwise. What is done here to address this is the addition of the rest of the original color tuple. pngimage.getpixel((x, y)) [3:] will have nothing in it if there are only three color values; if we have a transparent image, it will have the one value corresponding to the transparency. Remember, this is a forgiving slice operation that we're using. So we will either have nothing in that slice (for an RGB image) or we'll have a single value, the alpha value from the original image.

Find a PNG image online that does not have transparency (these may not be as common, but a search can turn one up) and try out this code to see that it will work on images whether or not they have transparency.

Now that you've written a function to lighten an image, copy the code, paste it in, change the function name to darken(), and modify the function so that it makes an image darker instead of lighter.

[10.8]   Increasing the Contrast of an Image

How should we change an image so that it's higher in contrast—so there is more difference between the light and dark areas?

Consider this: You've already done everything necessary in terms of image modification. You saw how to lighten an image and how to darken it. Now, all you need to do is make the lighter areas (those closer to white) even lighter and the darker areas of the image (those closer to black) even darker.

The only question is how to determine which are the light pixels and which the dark. Each pixel has three color values, and if the pixel is completely dark—is black—they are all zero. If it's white, they are all 255. The *sum* of these values is $0 + 0 + 0 = 0$ for black and $255 + 255 + 255 = 765$ for white. If we want to find pixels that are closer to black or closer to white, we would do well to sum their color values and see whether they are closer to 0 or closer to 765. The dividing line is 765 / 2, or 382.5. This will be the essential part of the *condition* we use to determine whether we want to darken or lighten.

Go back and retrieve the generalized `modify()` code, which works in both RGB and RGBA mode, and paste it in, renaming `modify()` to `contrast()`. Then, replace the line where `new_color` is updated:

```
new_color = (r + 64, g + 64, b + 64)
```

with the following:

```
if r + g + b < 382.5:
    new_color = (r - 32, g - 32, b - 32)
else:
    new_color = (r + 32, g + 32, b + 32)
```

The condition tests whether the color values are, overall, more light or more dark. If the pixel is more dark, our contrast-enhancing function makes the pixel *even darker*. Otherwise, it makes the pixel lighter.

Try it out: the result, even if it's not as slick as what an image manipulation application provides, should exhibit the increased contrast that was sought. This is the essence of a contrast-increasing image modification.

[10.9]   Flipping an Image

Next, let's write a new method, `flip()`, which performs the Flip Horizontally action found in Photoshop and other image editors.

To flip an image horizontally, or any other way, we obviously need to read the pixels of the image, and we certainly need to write them as well. Consider a *very* simple image, two pixels wide and one pixel high. Of course, we just want to put the right pixel on the left and the left pixel on the right. But let's think about how we might do that. If we read the right pixel, write it onto the left side, and then . . . we've wiped out the left pixel, which we're going to need in the next step. What needs to be done here, and in flipping an image in general, is a classic case of the *swap*, exchanging values between two variables.

Try this:

```
a = 17

b = 2↵
```

How do we put what is in a in b and what is in b in a? There are actually some clever ways and nice ways to make the exchange in Python, but the general principle is to use a temporary variable—for instance, temp:

```
temp = a

a = b

b = temp
```

Check to see that the exchange was made:

```
(a, b)
```

Having tried this swap technique on two values, let's swap every pixel (going left to right) with the corresponding pixel on the other side (going right to left). Before, we used (r, g, b) to get each color value component of a pixel. Now, we can simply put the entire triple in a single variable because we aren't going to manipulate the individual values. Also, the following function flips the way that iteration is done. The outer loop is y (vertical) and the inner loop is x (horizontal). It really makes no difference, in terms of how the function operates, but it seems clearer at this point to present this type of iteration. Finally, the following function uses *two* temporary variables—again, not because it is required, but in the interest of presenting code that is as clear as possible:

```
def flip(pngimage):

    width = pngimage.size[0]

    for y in range(pngimage.size[1]):

        for x in range(width):

            left = pngimage.getpixel((x, y))

            right = pngimage.getpixel((width - 1 - x, y))

            pngimage.putpixel((width - 1 - x, y), left)

            pngimage.putpixel((x, y), right)
```

Consider that width is set to the horizontal size of the image. Then, the horizontal loop runs from 0 up to (but not including) width. Each time, left is assigned the pixel (x, y) and right is assigned the pixel on the other side: width - 1 - x. Notice that there is no pixel at position width; the rightmost pixel is one to the left of that position. Hence, when x is zero, width - 1 - x or simply width - 1 is the rightmost pixel. The calls to putpixel() swap out the left and right pixel.

Obtain an image and place it in the directory where Jupyter Notebook is running, calling it `flipme.png`. Be sure that it isn't horizontally symmetric, or even close, so you can easily tell when it is correctly flipped. Then, run the function and see what happens:

```
testimage = Image.open('flipme.png')

flip(testimage)

testimage↵
```

Does it work? Let's get on to the real question: Why doesn't it work? This function doesn't appear to change the image in any way.

Actually, it *does* work. It just works too well.

As the function iterates from left to right within `for x in range(width)`, it avidly exchanges each left-hand pixel with each right-hand one. It does that for the first pixel on the left and right, the second pixel on the left and right, and so on. By the time it reaches the *middle* of the image, the left side and right side have been completely swapped.

Then, as the iteration continues, the function keeps swapping. The pixel just to the right of the center is exchanged with the one to the left. Eventually the rightmost pixel in the image is exchanged with the leftmost. Everything gets swapped back into its original place.

The function, as written, keeps going and undoes the desired swap that it has done to begin with. If you recall, this is the same sort of action taken by the iterative palindrome checker that you wrote in 8.10, Verifying Palindromes with Iteration and Recursion. In palindrome checking, it just ended up verifying the same thing twice or detecting the same mismatch twice, so while it was redundant, it didn't cause the function to work improperly. In flipping an image, going past the midpoint undoes the work that was properly done in the first half of the process.

So, what's needed isn't additional work, but less work. Instead of iterating over `range(width)`, how about iterating over *half* of that interval?

Give it a try and confirm that you have a working horizontal flip algorithm.

If you find that you have an error, you might be determining something like 200.0, or even 122.5, as the midpoint of your image. You can use `//` (integer division) or you can cast the result of the division so that it becomes an integer, using `int()`.

To conclude, I'll mention that Python has a handy built-in way to swap values using multivariable assignment. The following line of code will exchange the values in the variables *a* and *b*:

```
a, b = b, a
```

Given that it's this easy to do a swap, why did I annoy you with the "manual" method of swapping values? When you use this multivariable assignment, Python undertakes the same sort of exchange that we've done internally. This syntax is a convenience—one that you are welcome to use, but not one that all programming languages provide. While I don't consider that every inner working of the computer is important for an exploratory programmer, understanding the straightforward way to exchange values is a simple and important concept in programming and was worth introducing.

[Exercise 10-2] Flipping along the Other Axis

Now, to be sure that you understand how this works, write your own `flop()` function, which flips the image vertically.

[Free Project 10-1] Cell-by-Cell Generators

Do this project two (2) times, once starting with a text grid and once starting with a generated image. Choose to do *either* a visual text generator (which will create a grid of characters) or a standard image generator. Your program should iterate through the text grid character by character or iterate through the image pixel by pixel. It should produce some grid of characters or some image, using any method you like. To begin, you could make every character the same or every pixel the same. After you start by doing this, you should elaborate your generator somehow. You can use randomness, interesting mathematical functions, or anything else. But essentially, go through and generate each character or each pixel. If your grid is going to be made out of text, you won't use PIL.

After completing a generator, adapt it such that if you made a text grid generator, you convert it to an image generator—or vice versa. Of course, you'll have to deal with some differences between a grid of characters and an image. The former doesn't have color, whereas the pixels of the latter don't have different shapes, the way that characters do. Nevertheless, see what you can accomplish by trying out this conversion.

[10.10] Essential Concepts

[Concept 10-1] Modules Can Be Used in Python

There isn't much difference between using a library or module, such as PIL, and using the built-in, standard facilities of Python. You should be aware of how to use `import` to provide access to the facilities of such add-on libraries, just as `import` is necessary to access some of standard Python (`random`, `re`). The way to create and use objects (such as Image objects) can be slightly different, too, and it's important to figure out how this works.

[Concept 10-2] You Can Ask Python Itself for Help

Using `help()` allows you to read Python's built-in documentation for different functions and methods. To see what all the available methods and attributes of a particular object are, use `dir()`.

[Concept 10-3] Images Are Rectangles of Pixels

An image, as typically represented in a computer, is a rectangle of pixels, each one of which, these days, has (at least) red, green, and blue values associated with it. The size of the image and the color value of each pixel is represented as a tuple, a data type that you should understand, and the overall model of an image should be clear to you at this point.

[Concept 10-4] Nested Iteration Goes through Every Pixel

When it's necessary to cover every element (every pixel) in a picture plane, one dimension of iteration isn't enough. What works for a list or string isn't going to cover a two-dimensional region. In this case, nested iteration (increasing the variable x a step at a time while going through all values of y) is a very useful approach. Understand how to iterate in two dimensions, as is necessary for analytical, transformative, and in many cases generative work with images.

[Concept 10-5] Values Can Be Swapped between Variables

You should understand how (using a computer program) you can exchange the values of two variables and should see how this applies in the case of flipping an image. Understand how to do this in Python using multivariable assignment, but also how to do it "manually" using the classic swapping pattern.

[11]   Image II: Pixels and Neighbors

[11.1]

We'll continue investigating image manipulations in this chapter. Having looked at the cases where each pixel is independent, and where the pixel on the opposite side is important, we'll next consider a type of image processing in which the local area around each pixel is important.

[11.2]   Blurring an Image

Consider what "blurring an image" means at the pixel level. If an image is *sharp*, its pixels are distinct from nearby pixels. Not everywhere, of course, but along lines that distinguish different regions in the image. If an image is *blurry*, that means that nearby pixels are similar in value—more similar than they would be in a sharp image, even if they are along such dividing lines.

So, one idea—and one that will serve very well—is that a blur() function should make neighboring pixels more similar.

Of course, it's possible to blur an image *more* or *less*, and it's even possible to blur it in different ways—but for now, let's see if we can do it at all. Performing a blur of any sort allows us to do another fundamental image processing operation and to understand what is behind this operation.

So, consider: If you take some more or less random pixel—say, from the center of an image—how would you make it more like its neighboring pixels?

You could check out the pixel above it, the pixel to the right, the pixel below it, the pixel to the left—and why not the diagonal pixels, too, which seem to be neighbors, even if they just touch the pixel at a corner. If the color values of these were, on average, darker than the current pixel, it would make sense to shift the current pixel to a darker color. If they were lighter, it would make sense to make the current pixel lighter.

So for each pixel, we're interested in the average brightness/darkness of the eight surrounding pixels. Each pixel has three color values, and there are eight of them; that's twenty-four values to add up. We *could* just write out an expression that adds up all twenty-four of these terms.

But such an expression, with so many terms, would seem to invite mistakes. So instead, let's use at least one type of iteration. We could iterate over the eight neighboring pixels or over the three color values. Clearly, the former type of iteration would result in more compact code. But consider what the coordinates are of the neighboring pixels of pixel (x, y). Starting in the upper left and going right, they would be $(x–1, y–1)$, $(x, y–1)$, $(x+1, y–1)$, $(x–1, y)$, and so on. Of course, they aren't all in a line, and so there isn't the most obvious structure to this list of eight points. So it might be easier to iterate through the three color values, which are just r, g, and b in the tuple (r, g, b). Specifically, if this tuple were assigned to the variable color, they are color[0], color[1], and color[2]. Those should be very easy to iterate through.

So having decided which way to iterate, let's consider what it takes to sum up a single color value, the red value (number 0), from all neighboring pixels. Then, we can generalize this to add up all three color values and average them all. To make this a bit more compact, but still fairly clear, let's call the image we're working with (previously known as ourimage) by the three-letter name pic. Assign an image to this variable using the PNG file you found online in the previous chapter. If you're opening up a new Jupyter Notebook, be sure to begin with from PIL import Image so that you'll have access to the Image object within PIL.

Summing the eight red color values can be done with the following expression with eight terms:

```
around = (pic.getpixel((x - 1, y - 1))[0] +
          pic.getpixel((x, y - 1))[0]+
          pic.getpixel((x + 1, y - 1))[0] +
          pic.getpixel((x - 1, y))[0] +
          pic.getpixel((x + 1, y))[0] +
          pic.getpixel((x - 1, y + 1))[0]+
          pic.getpixel((x, y + 1))[0]+
          pic.getpixel((x + 1, y + 1))[0])
```

I surrounded these eight terms with parenthesis so the whole long expression can be broken across several displayed lines. In Python, it's okay to take a parenthesized arithmetic expression or a list or a tuple and extend it over several lines. You could actually also put all of this expression into a single line; it would work just fine. However, it would be much harder to read, check, and edit in that case.

This is a pretty formidable expression. It's a good thing we won't need to write one three times as long. Check through the eight coordinates to make sure you have typed in all of them correctly. They are all the combinations of x - 1, x, and x + 1 in the first place and y - 1, y, and y + 1 in the second place—*except* for (x, y), because we're only interested in the pixel's neighbors, not the pixel itself.

If we want to know the *average* red value, we can simply divide the value in around by 8 by adding this to the end:

```
around = around / 8
```

We haven't bundled this code into a function yet, but we can still test it out. Find a point in the interior of your image that is fairly dark. You can use an image editor to find the coordinates of a precise point on the image, or if you have an image that has a large dark region in the center or in one corner, you can figure out some point that will be in this region. For me, I set x and y as follows:

```
x, y = (127, 400)
```

Then, I pasted in and executed the long assignment beginning with around = pic.getpixel(and I also executed around = around / 8. That left me with an around value of 40.75, which is fairly close to black. I then located a region that looked white and found the coordinates of a pixel in that region. Again, you can use an image editor to find a precise location, or you can take a guess and check the color value of the pixel you picked. For instance, if you want to see the color values of the pixel at (200, 170), you can call pic.getpixel((200, 170)) and see what results. I did, and it was a good pixel to pick, so I set the values of x and y accordingly:

```
pic.getpixel((200, 170))↵

x = 200

y = 170↵
```

Again, I ran the expression beginning with around = pic.getpixel(, and I also ran around = around / 8. This time, around had the value 255.0, which means that all eight of the nearby pixels had the maximum red value, which they would if they were completely white.

This type of testing doesn't guarantee that our expression is the correct one; we could have included one coordinate twice, for instance. But it is useful to see that no error message is produced. Or, if one is, it's useful to fix the expression at this stage, before even more complexity is added.

Next, let's elaborate the expression to go through `color[1]` and `color[2]` in addition to the red component:

```
around = 0

for c in [0, 1, 2]:

    around = (around +

            pic.getpixel((x - 1, y - 1))[c] +

            pic.getpixel((x, y - 1))[c] +

            pic.getpixel((x + 1, y - 1))[c] +

            pic.getpixel((x - 1, y))[c] +

            pic.getpixel((x + 1, y))[c] +

            pic.getpixel((x - 1, y + 1))[c] +

            pic.getpixel((x, y + 1))[c] +

            pic.getpixel((x + 1, y + 1))[c])
```

We now need to accumulate the values for each color component in `around`. So we start by setting `around` to 0 and then go on to *add* each of the eight color values to it each time.

We also need to iterate through the three colors. We can do that without trouble by just iterating through `[0, 1, 2]`. We could have written `range(3)` instead, but it wouldn't really have been clearer and it wouldn't have saved much space, either.

With those changes made, there is only one more needed. Where we looked for element number 0 with `[0]` in the expression, we now look for element number c with `[c]`.

The final change is that our last line, dividing `around` by 8, should be modified to divide by 24:

```
around = around / 24
```

We haven't done any blurring yet; we've just checked to see how light or dark neighboring pixels are. Still, let's bundle up what we have in a function so we can use it later. Now that we are making a general function, instead of doing a specific operation on our particular image called `pic`, I am going to use a new name for the image, which is an argument to the function. Instead of `pic` (one particular image), I will let this general argument be named `img`:

```
def nearby(img, point):

    (x, y) = point

    around = 0

    for c in [0, 1, 2]:

        around = (around +

                    img.getpixel((x - 1, y - 1))[c] +

                    img.getpixel((x, y - 1))[c] +

                    img.getpixel((x + 1, y - 1))[c] +

                    img.getpixel((x - 1, y))[c] +

                    img.getpixel((x + 1, y))[c] +

                    img.getpixel((x - 1, y + 1))[c] +

                    img.getpixel((x, y + 1))[c] +

                    img.getpixel((x + 1, y + 1))[c])

    return around / 24↵
```

nearby() takes two arguments. The first is the image it will examine. The second is a tuple (specifically, a pair) with the horizontal and vertical coordinates of a particular point in that image, a particular pixel.

Try it out by loading up a new image using the Image.open() method. Then assign that to the variable ourimage, and check a few points inside that image—for example:

```
nearby(ourimage, (42, 40))↵
```

One last modification to this code, which we'll rename as differ(). Let's have our function determine not only the average nearby color values, but also the *difference* between that average value and the current pixel's value. We can do this by also adding up the color values of the current pixel and using them at the end. If the surrounding area is darker (closer to 0), we should find that this function returns a negative number. If it is brighter (closer to 255), the function should return a positive number.

```
def differ(img, point):

    current = 0

    around = 0

    (x, y) = point

    for c in [0, 1, 2]:
```

```
current = current + img.getpixel((x, y))[c]

around = (around +

        img.getpixel((x - 1, y - 1))[c] +

        img.getpixel((x, y - 1))[c] +

        img.getpixel((x + 1, y - 1))[c] +

        img.getpixel((x - 1, y))[c] +

        img.getpixel((x + 1, y))[c] +

        img.getpixel((x - 1, y + 1))[c] +

        img.getpixel((x, y + 1))[c] +

        img.getpixel((x + 1, y + 1))[c])

return (around / 24) - (current / 3)
```

Make sure you understand what this elaboration is and why it was done. We could have obtained the proper value for current differently. It would have been possible to simply add up the color values in the current pixel with an expression of only three terms. But because we are *already* iterating over the three colors, I wrote the preceding code, which accumulates the color values in current using that preexisting loop. I don't know that it's more concise or even that it's the best possible option, but it is at least consistent with the way the value of around is being determined.

Try out differ() to see if it provides reasonable results. You may not be able to locate "sharp" areas of an image to target, but you can at least make sure that you are getting reasonable results—never less than -255 or more than 255, for instance.

Now we're finally ready to blur images. We need a function blur() that uses the framework we developed earlier to visit every pixel in an image. For every pixel, call differ() and get a number showing how much this pixel varies from the surrounding ones. In continuous, similar-looking regions, this should be near zero. In "sharp" regions, it will have a larger magnitude, and may be negative (if the pixel is lighter than average) or positive (if darker). To blur the image a bit, we can shift each color by half of what differ() returns.

One more thing that's important to note at this point. When we were going around and changing individual pixels based only on those individual pixels—as when we were increasing the contrast, for instance—it was no problem to modify the image as we read it. But now that our new pixel depends on *other* pixels around it, we'd better create a new image and return that one. If we try to modify the image in place, our changes will start influencing future changes that we'll make, and we won't get the right results.

In fact, this modification-in-place problem was what we already experienced with flip(). In that case, the problem was solved by just making a copy of *one pixel* as we proceeded down the image. But we can't get away with that in blur().

So, we'll create a new image of the same size as ourimage by calling Image. new('RGBA', img.size). Call it newimage. You may notice that we're not specifying the color. It's optional and is set to (completely transparent) black if we don't specify a value. Because we will visit every pixel and set the color for each one, there's no need to set the color to anything in particular initially.

So, here's the first thing to try:

```
def blur(img):

    blurred = Image.new('RGBA', img.size)

    for x in range(img.size[0]):

        for y in range(img.size[1]):

            (r, g, b, a) = img.getpixel((x, y))

            change = differ(img, (x, y)) / 2

            blurred.putpixel((x, y), (r + change, g + change, b + change, a))

    return blurred
```

This code iterates along the width and within that the height of the image, and for each pixel it obtains the color values. It determines half the difference from the neighbors, and then sets the pixel's red, green, and blue values appropriately. The original alpha value of each pixel is preserved.

However, it has some problems. You'll see the first of them when you run this code—for instance, on ourimage—and witness TypeError: integer argument expected, got float, with an indication that the problem is in putpixel(). The color values being put into each pixel of the image have to be integer values. One easy way to fix this is to explicitly cast change so that it becomes an integer. To be sure you understand how this process of *casting*, this conversion from one type to another, is done, try out the following:

```
int(500.2)
```

```
ratio = 120.7
```

```
int(ratio)
```

The use of int() *truncates* the digits after the decimal point while it changes the *type* of the data from floating point to integer.

Let's do something slightly different and round to the nearest integer, using another built-in function, round():

```
round(ratio)↵
```

That changes the type as well. Check by trying out type(round(ratio)).

So, replace differ(img, (x, y)) / 2 with round(differ(img, (x, y)) / 2) on the sixth line of blur(). This way, each of the integer color values r, g, and b will have an integer added to them, and the result will be an integer. The alpha value will remain as it originally was.

After fixing this, the next error you should see is IndexError: image index out of range. The error message also shows that the problem occurs within differ(), on the sixth line of blur() and the fourteenth line of differ(). Let's start at the very first step this function will take to try to determine what is causing this error.

At the first step of iteration, x is 0 and y is 0, corresponding to the point in the image that is in the upper left. Getting the color values of pixel (0, 0) shouldn't be a problem. Indeed, the error is not happening there. In the next line, where the value of change is determined, there is a call to differ(). This function iterates through color 0, 1, and 2 and for each one adds up the color values of the neighboring pixels. So specifically, in this first step, it will first add up the color 0 (red) values for the neighboring pixels of (0, 0): that is, pixel (-1, -1), (-1, 0), (-1, 1), (0, -1), and so on.

This seems a little confusing. What could pixel (-1, -1) mean? As it happens, *now* (in the current version of Python and PIL) this coordinate indicates the pixel in the lower-right corner. That is, for an image 200 pixels wide and 100 pixels high, the upper-left pixel is (0, 0), and (-1, -1) is the same as pixel (199, 99). This is consistent with the way you can use -1 in indexing and slicing strings to refer to the last character of the string, as described in 8.10, Verifying Palindromes with Iteration and Recursion. So the beginning of the process, at the start of this first row, isn't causing a problem. But what happens when we get to the end of the row? Let's be concrete and consider what happens for an image 200 pixels wide and 100 pixels high. At the last step of the first row, we'll be checking (-1, 198), (-1, 199), (-1, 200)—and *there* is our problem. Even though Python and PIL have been updated (since the first edition of this book) so that (-1, -1) now refers to a pixel, (-1, 200) isn't going to refer to anything. We can only go up to 199. Hence the error that told us index out of range. This should be familiar! It's the same sort of problem we encountered in 8.4, Counting Double Letters.

A fix (good enough for this exercise) is easy to apply. Instead of beginning our horizontal iteration at 0 and going to width - 1, we should only go up to width - 2. For consistency's sake, let's also start at 1 so that we don't begin or end on the edge of the image. And we should do similarly for our vertical iteration. In other words, we will only attempt to change the *interior* points. We'll just leave the outer-border pixels as they originally were.

```python
def blur(img):
    blurred = Image.new('RGBA', img.size)
    for x in range(1, img.size[0]  - 1):
        for y in range(1, img.size[1]  - 1):
            (r, g, b, a) = img.getpixel((x, y))
            change = round(differ(img, (x, y)) / 2)
            blurred.putpixel((x, y), (r + change, g + change, b + change), a)
    return blurred↵
```

[Figure 11-1]
Interior pixels have eight neighbors, the way we've defined neighbors—including those pixels on the diagonal. But pixels on the edges and in the corners don't, so code that tries to refer to all eight of these neighbors won't work.

At this point, we have developed a correct function to blur an image in one particular way. This function doesn't look very far away from each pixel it is adjusting when it blurs the image; it looks only one pixel away from each pixel. But it does manage to blur the image at this low level. If you can't see the blurring that has happened on the image you chose, construct an image that has a single, one-pixel line. (Any image-editing program—even a paint-style program—will allow you to create such an image.) Blur that image, and then take a look at it, zooming in to see what has happened. You'll notice that the blur() function does indeed blur the image, although only in a very contained way.

To make the effects of blurring more visible, you can apply your blur() function multiple times, taking the image you have in memory (let's say it's stored in the picture variable) and trying out something like blurry = blur(blur(picture)), then inspecting the result.

You can imagine, I'm sure, how to blur an image like this with a wider window for each pixel. Instead of looking only at the eight neighboring pixels, you could look at the sixteen pixels around those.

As with flipping an image, this function performs the work of blurring an image just as Photoshop, the Glimpse editor, and other programs do. Image-manipulation programs are more general, but the underlying principle is the same one seen here. In fact, blur() is a specific case of a more general technique in image manipulation—the use of *filters*.

Blurring an image in this way is a typical exercise in an introduction to programming. You can imagine that the problem could be solved by a beginning programmer who has developed some intuitions, understands how iteration works, and can easily deal with arithmetic. In this section of the book, I've tried to explain how the step-by-step development process might proceed and what difficulties might crop up. These are very typical difficulties for this type of assignment, and it would be valuable to encounter them and overcome them on your own.

Independent learners don't have the benefit of working through a problem like this one and overcoming every difficulty, however. Even classroom learners who collaborate with other students don't always complete such problems with good understanding. In an ideal world, they would work together and discuss each problem posed, collaborating through a process like this one.

I hope that you will encounter several interesting and instructive difficulties, and overcome them, in this book's free projects. In this section, I have modeled a programming process, complete with particular ways a programmer might encounter errors and (with the help of error messages) figure out how to fix the code. In typing in these snippets of code, you of course have the opportunity to unwittingly introduce errors that you will also have to address, so there will be some opportunity to go over the code closely and deal with difficulties that I haven't documented.

[11.3] Visiting Every Pixel

I mentioned previously that to accomplish global image manipulations, our program needs to visit every pixel of an image, at least to read the color values of each pixel. At this stage, I will acknowledge that this statement, while it provides reasonably high-level guidance, is not completely correct. Of all the image manipulations we considered, there is one type that does not *always* require that the program visit every pixel of the image. Although this is a fine point, it presents a good opportunity to test and improve your thinking about programs and how they work.

Which of the image manipulations we covered does not, in absolutely every case, require that every pixel be visited? In what cases must every pixel be read and/or written, and in what case can some pixels be ignored?

If you can answer these questions now, go ahead and do so. If not, read on for an example of how to prove that, for a particular image manipulation, the program must visit every pixel.

Consider the manipulation called *de-red*, which is sort of an extreme version of the cooling manipulation. This sets the *red* color value of every pixel to 0. In this case, there are two ways that the program might operate.

The program could read the value of every pixel. If the red value is already 0, no action is required for that pixel. Otherwise, it must set the pixel with 0 as the red value. In this case, even if some of the pixels have no red to begin with, every pixel still must be visited in order to *read* it and determine this.

Or, if there is a way to set the red, green, and blue color values independently, the program could simply set the red value of every pixel to 0 without even bothering to read it. But in this case, every pixel must be visited in order to *write* each one of them.

There is an even simpler way of putting it that doesn't get into the details of reading and writing: If the program *doesn't* visit some pixel, that pixel might need to be manipulated to remove the red, and if it does need to be changed, the program will not be able to complete the de-red operation. So it must visit every one. There is simply no shortcut that can be guaranteed to work with all images.

Apply this reasoning carefully to the image manipulations we implemented. In what specific case, for what manipulation, does this argument not hold? It shouldn't be obvious, but go through each of the manipulations we've done. Is there one that might succeed without reading or writing certain pixels? Because we know, by virtue of how the image manipulation is defined, that some pixels are correct as is?

Here's a hint: Consider the case where an image's height or width is odd.

[11.4] Inverting Images

Let's develop a new transformation, although a very simple one. We're going to *invert* the image, developing a function invert(). White will become black and black white. If a particular pixel has the maximum red value (255) and no green or blue (0 in both cases), we'll change that so that the new pixel has no red (0) and has the maximum green and blue (255 in both cases). If we have a pixel that is very dark gray—for instance, (20, 20, 20)—it will become very light gray: (235, 235, 235).

First, let's figure out how to iterate through the image. When we wrote blur(), it was necessary to look at the surrounding pixels each time we wanted to adjust a pixel. We also made a copy of the image so that our transformation didn't influence itself along the way. Is that necessary for invert()? Is the way we change a pixel defined in terms of its neighbors or in terms of the pixel itself?

Having determined that, select a framework. If we are going to change the image a pixel at a time based only on the previous value of the pixel, you can start with modify() and rename it invert(). If we need to check the nearby pixels to determine their values, it would be better to start with blur() and differ() and modify those.

As we iterate through the pixels, we need to check the current color values of each one. The getpixel((x, y)) method, a method of the Image object, allows us to do this. Its output can be assigned to the same sort of triple we've been using in previous problems.

Finally, we need to write a new pixel with new color values. The putpixel((x, y), (r, g, b)) method, also a method of the Image object, allows us to do this. But instead of writing the values (r, g, b), we want to write the opposite of r, the opposite of g, and the opposite of b. If one of these values is 0, we would like 255. If one of them is 20, 235. If one of them is 127, 128. And so on. What expression can be used instead of r to produce this result? It's a short expression indeed, one that has only two terms. Once you determine what it is, you should write similar expressions for g and b.

Hopefully that was neither too plodding nor too cryptic. At this point in this chapter, you should be able to write the invert() function. Give it a try, and if entering it or testing it out produces errors, work to correct them.

If you don't understand how to write code to invert an image, you should identify what question or questions you have. You should understand the following:

- How to do two-dimensional iteration through an image
- Whether you need to check the neighboring pixels or simply consult each pixel individually
- Whether you need to make a copy of an image or can simply write the new color values in place to each pixel along the way
- How to get a pixel's color values and store them in three variables
- How to write new color values to a pixel
- How to invert a color value so that, for instance, 0 becomes 255

In all but the last case, if you have a question, you should review the discussion earlier in the chapter, entering and modifying code, to better understand these topics.

In the last case, keep trying out different expressions, using the Python interpreter as a calculator.

Finally, when you have written invert() successfully, look over that list to see some of the specifics you have learned about. To those, I'll add that we also have discussed, and a reader/programmer at this point should be conversant with:

- How bitmapped images are represented at a low level and why their color values range from 0 to 255
- How to load images from and store images to the file system from within Python, where they can be manipulated via the Image object
- How the contrast of an image can be increased
- How a particular filter (of the sort used in Photoshop and the Glimpse editor) actually works at a low level
- How to convert between types (floating point and integer) and why one would need to convert
- How saturation arithmetic works, and how it makes setting the color values of pixels easier

You should really feel ready, at this point, to jump in immediately and start writing code that modifies images and analyzes them. You might want more practice with image manipulations and with reading images, of course, and you might want to learn some higher-level and more practical techniques for working with images. The work done in this chapter is the basis for such further exploration.

[Exercise 11-1]   Old-School Filter

Here's another type of image manipulation to try. Convert ordinary pictures to make them look (sort of) as if they are being shown on an old monochrome, green-on-black monitor. To start with, you need only change the color; this manipulation doesn't have to add scan lines, phosphor patterns, or blur—although of course the blur manipulation could be used to do the last of these.

[11.5]   Practical Python and ImageMagick Manipulations

There is a simpler way to invert an image using PIL. After importing the Image object, if it hasn't been done already, and also importing another object as shown on the second line ahead, an image can be inverted with a single command. Let's look at the whole process of doing the needed imports (as if in a fresh Jupyter Notebook), loading a PNG image from a file, converting that image to RGB mode (the conversion will just do nothing if it is in RGB mode to begin with), inverting the image, saving the inverted image, and then displaying it in the notebook:

```
from PIL import Image

import PIL.ImageOps

ourimage = Image.open('source.png')

rgbimage = ourimage.convert('RGB')

inverted = PIL.ImageOps.invert(rgbimage)

inverted.save('inverted.png')

inverted↵
```

For well-known transformations (such as flipping, blurring, and inverting), it is best to use built-in methods such as these. When they are part of a reliable package such as PIL, they will tend to be well-written and well-tested. We didn't care about how fast our code was, and for custom research purposes, optimizing for speed often doesn't make sense. But when one is doing a large number of image manipulations, particularly in an automated and regularly repeating way, it can be a big help to use code that is so optimized.

There are several standard image manipulations you would want to use in processing a batch of images for the Web or for another project. Although it's nice to learn how these are done at a low level, and in some cases one may wish to program new image manipulations, a great deal of common image manipulation work can be done with existing programs and without using Python and PIL. Photoshop and the Glimpse editor are set up to manipulate individual images, but there are always ways to go through large numbers of images and process each of them in the same way. A great way to do this is using ImageMagick, which is free software and works across platforms.

ImageMagick can be used to convert between image formats in addition to manipulating the pixels of images.

ImageMagick is also frequently used as a command-line program, and the ability to work in the shell (which you are developing or at least practicing in this book) will help you to use ImageMagick. This isn't the place for a complete tutorial, but here is an example of how it works:

```
$ convert -negate *.png inverted.png
```

For each PNG image in the current directory, this creates a file (named inverted-0.png, inverted-1.png, etc.) that contains the inverted (or *negated*) image.

Using wildcards such as *, the command line provides simple ways to refer to large numbers of files at once. The command-line arguments to ImageMagick must be learned, or the documentation must be consulted to determine how to compose them, but they provide a concise way to specify the operations that are to be done and the order in which they should take place. This is not to say that ImageMagick is *better* overall than a GUI application, simply that it has advantages when there is a batch of images to process; it is good at generalizing across images. For making manual pixel-by-pixel changes, these graphical applications are almost certainly better. But for applying a few standard manipulations, and possibly a format conversion, to a large number of images, ImageMagick works well. Just as textually writing programs provides for a good way to generalize over numerous images, the ability to type command-line arguments also helps people as they specify what ImageMagick is to do to large numbers of files.

Finally, because one can invoke ImageMagick from shell scripts and programs, it is possible to automate its workings and integrate it into a larger systems, such as a website. The textual interface, which users of the program can access directly by typing on the command line, also serves as an application programming interface (API) and allows *programs* to run ImageMagick.

Of course, there is a major limitation to *only* knowing how to use ImageMagick, powerful as it may be. If you devise some sort of transformation or method of analysis that is not supported by this package, you'll either need to code it from scratch or forget about it. As your transformations become more nonstandard, more unusual, you may also find that they are *possible* to do with ImageMagick but are *easier* to do in Python.

[11.6]   Manipulating Many Images

Going through every pixel of an image, for every pixel in a row and for every row in the image, is a type of abstraction and generalization. What can be done for one pixel can be done for every pixel in an image. In this section, we'll generalize our image manipulations further and show that what can be done for a single image can be done for any number of images.

The practical uses of such manipulations include preparing a body of images for a digital media art project or for a website, in ways that are more complex or specialized than can easily be done with ImageMagick.

We've seen how the command-line program ImageMagick can work on every PNG image in a directory when given the argument *.png. Wouldn't it be great to have our Python image manipulations work on every file in a directory as well? This would allow us to generalize an operation we can do to one image and have it apply to many of them.

Let's use the old-school filter you wrote in the previous exercise. If you haven't developed the filter in this way already, modify it so that it is named old_skool() and accepts an Image object (not a filename) as an argument. I'm assuming that your function modifies the image it is given "in place" and does not need to return a new image. If you do return a new image, you'll have to change the example code ahead.

We'll write a loop that wraps around the call to old_skool() and for every file in the directory opens it, processes it, and writes the modified file. In the middle of our loop, we'll have:

```
modified = old_skool(current)
```

In that line, the filter will be applied to the current image and will result in the processed image, stored in a new variable. But with just that line, we won't have the image in memory yet, and when we're done we won't have saved it. We also need a line for each of those. Let's first write our program to process a single file, example.png, and then go back and write code to generalize this program:

```
current = Image.open('example.png')

modified = old_skool(current)

modified.save('example.png')
```

Wait before pressing Shift-Enter. There's something we want to edit here. This code will wipe out the original example.png and replace that file with the modified image. If that's what you really want to do at some point, okay, but often you don't want to overwrite your original file. And when you're still making sure your program works properly, it's a particularly bad idea! Let's make a minor change, just to line three, so we save a file with a different name and don't overwrite the original image:

```
modified.save('oldskool_example.png')
```

This isn't perfectly safe! If you have a file named `oldskool_example.png` already in the directory, that will get overwritten. But let's make this improvement, for now, and go on. Try these three lines with the appropriate PNG file in the directory and see that the transformation works for `example.png`.

Now let's figure out how to do this for every PNG in the current directory. To start, we can get a list of what files ending in ".png" are in that directory. There are a few ways to do this, but it's easy to use the `glob` module:

```
import glob

glob.glob('*.png')↵
```

That provides a list of all files in the current directory with the appropriate file extension, using a syntax that is very much like our command-line example. It's a list like any other list, so we can iterate through it and work on each filename:

```
import glob

file_list = glob.glob('*.png')

for filename in file_list:

    source = Image.open(filename)

    modified = old_skool(source)

    modified.save('oldskool_' + filename)
```

That's all there is to this type of generalization—taking some computation that we're performing on a file and performing it on many files. Notice that if you have *n* different PNG files in your directory, this always places *n* new files, with modified images, in that same directory—even if some of the original filenames already begin with `oldskool_`! In that cacse, you'll get a filename beginning with `oldskool_oldskool_`.

Of course, `glob` can be used to work with directories and files in more powerful ways, and there are also some limitations to this approach. If we have a file in our directory that ends in ".png" but isn't actually a PNG image, that's almost certainly going to crash the program we wrote. We could revise the program to account for that and to produce an error and continue in this case. But this is a pretty good starting point for multifile image manipulation.

[Free Project 11-1]   Image Manipulation as You Like It

Do this project at least two (2) times. Select an image transformation that is available in some existing software, one that you are curious about. You may choose Photoshop,

Glimpse, a bitmapped image editor in the style of Microsoft Paint, or even ImageMagick. Implement this transformation in the same way we implemented flip(), blur(), invert(), and other manipulations. You may find, after trying for a while to implement your selected transformation, that it is proving too challenging. In that case, just choose another one that seems more suitable. The time you spend working on a transformation that turns out to be complex isn't wasted; it's part of learning about what computational work is more straightforward and what work is harder to implement.

[11.7]   Essential Concepts

[Concept 11-1]   Checking the Neighborhood
By developing a function to blur an image, you explored how the new value of a pixel can depend on nearby pixels. Computation that looks to the local area is applicable to analysis as well as transformation and applies to other media. Understand the basics of checking nearby values.

[Concept 11-2]   Working One Step at a Time
After you have an immediate goal and know the approach you want to take a programmer, develop the first step of this approach and test it out as best as you can. Address the first error you encounter, then the next.

[Concept 11-3]   Generalizing to Many Files
What can be done with a single file can be done to a large number of files, whether it's image manipulation, the processing of text documents, or any other sort of data. Understand how to expand a program written to work on one file to work with (at least) every applicable file in a directory.

[Concept 11-4]   Customizing beyond What's Standard
With image manipulation and many other aspects of programming, any capable programming language will offer the ability, either built-in or using libraries such as PIL, to do a variety of standard types of operations. You should use these when you want to do any standard type of operation because they will be optimized and well-tested. When you want to do something that isn't standard, however, you'll benefit from knowing how computer programs work with images and other media at a low level. Being able to program more generally, whether for image manipulation and analysis or for other purposes, will allow you to pursue explorations off the beaten path.

[12]   Statistics, Probability, and Visualization

[12.1]

In this chapter, three important and related ways to understand data will be introduced both through practical experience with programming and through theoretical discussion. The first of these approaches is *statistics*, a systematic, mathematically principled methodology for analyzing quantitative data, which relates directly to the second approach, *probability*, which offers a perspective on the underlying processes by which such data may be generated. The other way to understand data is through *visualization*, a designed way of visually representing quantitative data to allow better exploration and understanding. Measures that are derived from data, such as averages—and, more specifically, the mean we implemented in 6.4, Iteration (Looping) Abstracts along Sequences—are also themselves called statistics. There are plenty of relationships among statistics, probability, and visualization: all are means of providing new perspectives on data, for instance. Something discerned in a visualization will often lead a researcher to do further statistical analysis.

There are various reasons to do statistical, probability, and visualization work. A researcher might have a rhetorical purpose and be trying to persuade others by showing statistics and visual representations that support one's point. This mode is different from that of the explorer who is trying to identify unexpected trends and correlations in the data. The presenter has a conclusion and is looking for ways to support it, while the explorer has data and is trying to figure out what can be concluded. I'm glad to have learned about this from one of my advisees, Michael Danziger, whose master's thesis explores this distinction and many other intriguing aspects of visualization as it exists in culture (Danziger 2008).

By noting that there is a difference between persuasion and exploration, I don't mean to say that the presentational or persuasive use of statistics and visualization is somehow wrong. In a community of researchers, each individual or team may use statistics and visualization persuasively when making a case for their findings, and by contesting and building upon previous results in this way, the community as a whole could certainly end up advancing toward better understanding. However, because our theme here is exploratory, the use of statistics and visualization to explore data will be our focus.

This work in visualization, and the work in animation that occupies the next chapter, can be done in Python but is very well-suited to another mature programming language, which is also free software: Processing. Rather than learn another optional Python module, I will introduce Processing in this chapter.

Processing is a streamlined, Java-like language that was developed originally so that visual artists could learn with the system and use it to sketch. Working in Processing is different in a few ways from programming in Python, but the fundamentals are actually the same. Shifting to Processing for a moment provides a way to see how our experience of programming is not just "learning Python," but can generalize to other programming languages. The next chapter, "Classification," involves working in Python again. This shouldn't be a jolt or cause a crisis, however. The same fundamentals of programming apply in both languages.

Processing is not really going to shine in this chapter the way it will in chapter 14, "Image III: Visual Design and Interactivity." This chapter will serve to introduce the language and show how some things we know how to do in Python can be accomplished in Processing. After working through this chapter, chapter 13, and chapter 14, you'll have a much better idea of how to get more deeply into visualization, along with more practical means to do so.

Processing also gives us the opportunity to do some programming in a different environment, a simple and effective IDE—which is not the same as an interpreter or a text editor, but, like other ways of programming, allows for code editing while providing other facilities. That means you'll have experienced programming in the exploratory environment of Jupyter Notebook, the plain old text editor, and at least this one IDE.

[12.2]   The Mean in Processing

Early on in this book, in 6.4, Iteration (Looping) Abstracts along Sequences, we determined how to find the *mean* of a list of numbers, using Python:

```
def mean(sequence):

    sum = 0

    for element in sequence:

        sum = sum + element

    return sum / len(sequence)
```

We also need to define some sequence of values in Python in order to determine the mean of that sequence. And we would need to call mean() with this sequence as its argument to actually compute the mean. So it would take more than these five lines to do the relevant computing work, although this is the core of it. The rest of the Python we need might look like this:

```
values = [10, 9, 73, 25, 33, 76, 52, 1, 35, 86]

print(mean(values))↵
```

In Jupyter Notebook, as you know by now, the print() function itself isn't necessary if you want to view the value because the notebook will report on what value is returned. To type programs into Processing, you open up Processing and type into the white pane that is below the menu bar and below the tab. Here's how to do the same computation of the mean, already accomplished in Python, in Processing:

```
float mean(float[] sequence) {

    float sum = 0;

    for (int i = 0; i < sequence.length; i = i + 1) {

        sum = sum + sequence[i];

    }

    return sum/sequence.length;

}

float[] values = {10, 9, 73, 25, 33, 76, 52, 1, 35, 86};

void setup() {

    println(mean(values));

}
```

Open up Processing and type the preceding code into your code window in the Processing IDE.

As with Python, the core of the work is done by the function mean(). Beyond that, you need to list the values (as you would in Python) and invoke the mean() function (as you would in Python).

Once you type this in and run it (using the triangular button that looks like a "play" button), you should see that in the console area, the black pane at the bottom, which is smaller by default, the number 40.0 will appear. That's the mean of the ten numbers stored in the variable values, and it's the same answer you will get if you type in all of the preceding Python code. As we start working with Processing, I'll mention the built-in Processing reference once again, as I did in 1.8, Programming and Exploring Together. You can get to this resource through the IDE's menu: Help > Reference. You can see there's an option to find terms in the Processing reference, too.

Let's consider the specific ways in which the Processing code differs from the Python code, starting with the way the mean() function is defined.

In Python, we begin with def mean(sequence):, but in Processing, we begin with float mean(float[] sequence) {. This is because the template for defining functions and methods is different in Processing. It looks like this:

```
TYPE ____(TYPE1 ____, TYPE2 ____, . . .) {

    ____

}
```

In Processing, it's necessary for the programmer to declare the type of each variable and what type of value a function or method will return. Without going any further into programming language design, I'll just mention that there is a rationale for this requirement. In this case, the function is supposed to return a floating-point number or float, while its one argument is supposed to be an array of floating-point numbers, which is indicated with float[]. An array is so similar to a Python list that it's not important to explain any more right now. That explains the first line; now on to the next.

Here we have float sum = 0;, which is different in two ways from the corresponding Python line. First, in this line of Processing code, we explicitly have stated that this is a floating-point number by starting with float. That's consistent with the way we used float and float[] in the previous line. Second, this line ends with a semicolon.

In Processing, as in Java, as in C, as in JavaScript, the end of a logical line of code must be explicitly indicated with a semicolon. It's not necessary in Python, and this has to do with the way Python uses indentation to indicate code blocks. In Processing, you can indent any way you like—although it's much clearer to do it in a way that is similar to Python and is easily understood visually. No matter what, though, each statement needs to end with a semicolon.

Next, consider:

```
for (int i = 0; i < sequence.length; i = i + 1) {
```

This line is also is different in some significant ways from the corresponding Python line. That's because the basic template for iteration, like the template for defining methods and functions, is different in Processing. Here's what it looks like:

```
for (INIT; TEST; UPDATE) { ____ }
```

The INIT or *initialization* is typically an assignment (in this case, int i = 0) that sets a *counter variable* (a variable that will accumulate value) to its initial value. It is run once at the beginning of the loop's execution. The TEST is the condition for exiting the loop. If the TEST is true (in Processing indicated with true and corresponding to the Python value True) the statements in the body of the for loop, whatever is in the ____, will be executed. In this case, i < sequence.length checks to see if the counter variable, i, is strictly less than the length of sequence. Finally, the UPDATE is typically used to change the counter variable, as it is in this case. Here, the update simply increments the counter variable i.

The way the length of the sequence is obtained here is not by calling a function, by checking len(sequence), as in Python, but by asking for the length *attribute* of sequence, by including sequence.length. That's just how the length is found in Processing; there are choices to be made in all programming language design, and in this case the designers of Processing (following Java) made a different decision than was made in designing Python.

In the next line, what our program is adding to sum each time is the current element of sequence. In this loop, however, it isn't held in the variable element but needs to be accessed as sequence[i].

This snippet of code then returns the mean. There shouldn't be many surprises waiting in this final line. Of course, the way the length of sequence is accessed is different from Python and consistent with the attribute access discussed previously. And of course, the line ends in a semicolon because we're now using Processing. But otherwise, it's the same.

Just two more paragraphs will cover the last two bits of code. float[] values = {10, 9, 73, 25, 33, 76, 52, 1, 35, 86}; is assigning a sequence (strictly speaking, in Processing, this is an *array*) of numbers to the variable values. Here we see this variable labeled with its type (float[]) and we see that the line ends in a semicolon.

Also, note the way an array literal is specified is with curly braces, { and }. It's just the recipe for assigning several values to an array when initializing that array in Processing.

Finally, we have some code to *print* (display to the screen) the mean of our values. One aspect of this code that is slightly unusual is that the function we use has a different name: println(). The Processing println() function just prints a newline (hence the *ln*) after whatever is to be output, however, so it shouldn't be too startling to see it used here. More unusual is that this println() is wrapped by something else:

```
void setup() {

  println(mean(values));

}
```

void is a special type for functions and methods that don't return any value at all. The function setup() is a special one in Processing, called when a program starts. When we get to more full-blown Processing programs, you'll see that it can be used to initialize some of the basic properties of the program, including graphical ones. For now, we're just using it to print a single number out to the console area.

This is a somewhat unusual way to introduce Processing because it doesn't take advantage of the impressive features of the language and its abilities to create visual sketches. We will get to those, though. This introduction does provide for a nice comparison between Processing and Python. You can see from this that there are some things to learn in starting to use a new language. But, also, you should be able to see that starting to program in Processing is not a revolutionary cognitive activity along the lines of "learning a new (human) language." It's just a matter of figuring out some differences of syntax—needing to declare types, needing to use semicolons, using a different function definition template, using a different template for iteration. Processing, like Python, has types, so that underlying concept is the same. Processing, like Python, allows iteration, so that underlying concept is the same. Processing, like Python, allows the encapsulation of code in functions and methods. The fundamentals do not differ here, just some surface aspects.

[12.3]   A First Visualization in Processing

Having written a Processing program that has some data packed into it, and that does the same computation of the mean that we accomplished earlier, let's now use a little bit of Processing's formidable capability for visual design. We'll simply elaborate the program created previously so that it draws a chart of the raw values and the computed mean.

To begin, let's figure out how we can draw anything at all—and let's do so by elaborating our existing program. We'll start by adding a line to the setup() function:

```
void setup() {

  size(200, 100);

  println(mean(values));

}
```

Try this version of the program and observe that the size of the output window (which is still not used for anything, at this point) is specified by the call to size(). Now, we were provided with an output window before we added this line. It was simply the default size of 100 × 100, which is pretty small. Even 200 × 100 isn't very big, but it will suffice for this example.

Next, let's add another line to show that we can draw a single rectangle using the rect() function. You have to know what the interface to this function is—what arguments this function takes—to use it, of course. The Processing reference allows you to easily look up how this function, and any other Processing function, works. Open it from the menu, Help > Reference, and under the heading "Shape" and subheading "2D Primitives," find the entry for rect(). You'll see that the basic format is rect(x, y, w, h): x is the x-coordinate of the upper-left corner, y is the y-coordinate of the upper-left corner, w is the width of the rectangle, and h is the height of the rectangle. So, see what a rectangle looks like with its upper-left corner at (30, 15) and its size specified as 20 wide and 10 high:

```
void setup() {

  size(200, 100);

  rect(30, 15, 20, 10);

  println(mean(values));

}
```

You may notice in the Processing reference that you can provide more than four arguments to rect(), allowing you to do fancy things like rounded corners. One way of understanding this is to consider that some arguments are optional, but you can also think of there being a four-argument function rect(), a five-argument function rect(), and so on. Processing is designed well so that these different functions work consistently with one another. Also, they are set up to allow for abstract thinking and a focus on the current project. You don't *have* to care about these other versions of rect(); you can use the four-argument version your entire life and be perfectly happy with just that one, if that's all that you need.

One thing you will notice is that the display window isn't set up like quadrant one of the typical Cartesian plane, the standard coordinate system you will be familiar with from visits to Mathland. Instead, (0, 0) is the upper left, which is why our rectangle appears in the upper left and not the lower left. You shouldn't be too startled to discover this as the Python Imaging Library works in exactly the same way. It's standard in computer graphics.

We've drawn one rectangle, but we really want to draw one for each data point, for each element of values. We can do this using iteration, by placing the call to rect() in a loop. For each data point, we'll call rect(). Here's one (incomplete) bit of code to do this:

```
void setup() {

  size(200, 100);

  //rect (30, 15, 20, 10);

  for (int i = 0; i < values.length; i = i + 1 ) {

    rect(__, __, __, __);

  }

  println(mean(values));

}
```

In this code, I've *commented out* our test rectangle by putting // at the beginning of that line. I could have removed this line entirely, but from the computer's standpoint, it's the same to comment the line out. This way, if we did want to display the test rectangle at some later point, we could do so by *uncommenting* the line—removing the //. It's possible to clutter things up by commenting out code, but if we just comment out a line or two it shouldn't be too much of a problem.

To run this, we need to figure out how to fill in the four blanks. Where should each rectangle—rectangle number *i*, corresponding to data point number *i*—be placed? How wide and tall should each one be?

To have these rectangles line up left to right, it would be sensible to make each one 20 pixels wide. We have ten data points and have made the display window 200 pixels wide, after all. If we wanted to make our plotting program *general* to all sorts of sequences, we could do a computation to determine the width of the window or the width of each rectangle based on how many data points we have. But we're going to get one specific case working first and then generalize later.

Now one of the four values has been determined: w will be 20. So what about x, y, and h? Let's put in some values, even though they won't be the right ones, to see what happens:

```
rect(0, 0, 20, 20);
```

This will draw a rectangle *each time* starting in the upper-left corner, (0, 0). Each of these rectangles will be 20 pixels wide and 20 tall. Think about what to expect from this code, and then run the code.

Did you expect to only see one rectangle? Of course, not knowing how rapidly Processing draws, you might have thought that you'd see a flicker as the rectangle in the upper left is drawn again and again. Except in some extraordinary circumstance, however, the drawing is much too rapid to observe. The effect is the same as if you had just drawn a single rectangle. This is not too exciting, but you can see that Processing is doing as instructed here.

Let's make one change, so that the x-coordinate increases as the program iterates through the data points:

```
rect(i, 0, 20, 20);
```

We've simply made the x-coordinate i, so that the first rectangle, number 0, is drawn at x-coordinate 0; the second, number 1, is drawn at (1, 0); the third is drawn at (2, 0), and so on.

Is the result consistent with your expectations? If not, can you see why the image looks like this?

Now it's possible to see that different rectangles are being drawn, one at each step, but things certainly don't look crystal clear. We really want the rectangles spaced out by 20, not by 1. A simple multiplication can accomplish this:

```
rect(i * 20, 0, 20, 20);
```

That puts the rectangles in the correct place (both horizontally and vertically, the upper-left points of each are positioned properly) and we already have the proper width set for each of them. So all that is left is to fix the last argument, the height. This measurement is supposed to correspond to the value of the data point itself. That is, rectangle i should have a height equal to values[i]. So let's put that in:

```
rect(i * 20, 0, 20, values[i]);
```

This is basically correct; there are only two things we should do to have this chart appear exactly as we had hoped. One thing we should do is add in something that shows the mean. The other fix, or at least optimization, would be to flip the chart so that zero is at the bottom and 100 at the top. It's not *wrong* as it is, but charts aren't usually set up this way, and it is confusing.

Right now, our upper-left point is anchored to the top of the window, while the height of the rectangle varies according to each data point. We *still* want the height of each rectangle to vary in exactly the same way, we *still* want the rectangles to all be 20 pixels wide, and we *still* want the rectangles to be 20 pixels apart. So of the four parameters of rect(), there is only one that we should change. We want to change the second argument, currently 0, to something that will place the rectangle with its base on the bottom of the window (100 pixels down).

Changing the 0 to 100 moves all the rectangles down and out of sight. Try it and see:

```
rect(i * 20, 100, 20, values[i]);
```

What we'd like to do is move them down not a full 100 pixels, but 100 pixels *less* the height of the rectangle. So if we have a rectangle that is 20 high, we place it at 80; if we have a rectangle 50 high, we place it at 50. We can have the program do this for us simply by using 100 - values[i] as that second argument:

```
rect(i * 20, 100 - values[i], 20, values[i]);
```

Here's what the setup() function should look like with this line:

```
void setup() {

  size(200, 100);

  // rect(30, 15, 20, 10);

  for (int i = 0; i < values.length; i = i + 1) {

    rect(i * 20, 100 - values[i], 20, values[i]);

  }

  println(mean(values));

}
```

There's really just one more thing to deal with. We have plotted all the data points, *and* our program is computing the mean of those points, but we don't have any graphical indication of where the mean is. The commented line (beginning with //) does nothing when the program runs; it can be there or not. The line with println() "prints out" the mean on Processing's console, but it doesn't do anything in the display window. What we would like to do is place an indicator of the mean on the chart.

There are a few ways to do this, but a simple one would be to draw a horizontal line where the mean is. If you check the Processing reference, you'll see that there is a line() function that works to draw lines in a very general way. But let me suggest that we can do something similar. Because we just want to draw a single horizontal line, we can think of this line as a very thin rectangle, perhaps with no height at all. We'll still see it because rectangles are drawn with lines that have some width. It just won't have any internal area to be filled in. We already know how to draw rectangles, so we don't even need to look up how line() works to deal with our immediate goal.

The upper-left point is going to be situated all the way on the left side of the window, so x should have the value 0. The rectangle will be 200 pixels wide (the entire width of the window) and, as we discussed, 0 high. So the only tricky thing is placing this line (this rectangle) at the right horizontal position. If we hadn't flipped everything horizontally, we could just use mean(values) as the y position. Because we have flipped everything, we need to use 100 - mean(values). Then, we should add a line like this one:

```
rect(0, 100 - mean(values), 200, 0);
```

The only question is, where should we add it? If we place it *before* the loop that draws the ten rectangles, it will be drawn "behind" those shapes. Try it out:

```
void setup() {

  size(200, 100);

  // rect(30, 15, 20, 10);

  rect(0, 100 - mean(values), 200, 0);

  for (int i = 0; i < values.length; i = i + 1) {

    rect(i * 20, 100 - values[i], 20, values[i]);

  }

  println(mean(values));

}
```

How does it look? I think it's not a complete atrocity, but it's also probably not the best visual display. Try placing the mean line (which is actually a rectangle) *after* this loop instead:

```
void setup() {

  size(200, 100);

  // rect(30, 15, 20, 10);

  for (int i = 0; i < values.length; i = i + 1) {

    rect(i * 20, 100 - values[i], 20, values[i]);

  }

  rect(0, 100 - mean(values), 200, 0);

  println(mean(values));

}
```

You may like this second version more or less, but in any case, you can see that you have options. You should see that it's also possible (although not very sensible) to put the call to rect() that draws this "mean line" *within* the loop, either before or after the other call to rect(). I consider that it isn't very sensible because conceptually, I don't think anyone would really want to draw this line in the same place ten times. It only needs to be drawn once.

After you've got this working, replace your use of rect() to draw the "mean line" with a call to line(). Figure out how to do this by consulting the Processing reference.

Here's what we have done so far, just in these first sections of this chapter:

- Computed the mean using Processing (having done it in Python already)
- Added code to Processing's setup() function
- Specified the size of the drawing window
- Drawn a rectangle
- Used the Processing reference to understand what arguments functions take
- Commented out code
- Plotted data points
- Flipped a plot vertically
- Added a line to indicate the mean

Now, if all you cared about was making a simple plot of this sort, you could of course use your favorite spreadsheet, such as LibreOffice Calc or Microsoft Excel. But you wouldn't learn much about how to computationally flip a graph vertically, how to comment out code, or how drawing different elements of a picture results in shapes being drawn on top of one another.

In the same way that you could use Glimpse or Photoshop to blur images instead of writing your own custom code to blur them, you can use standard programs to plot charts and graphs—and you should, when it's appropriate. But if you're going to develop an artistic or humanistic project that uses data visualization in new ways, you would like, of course, to be able to write programs that visualize data in new ways, not simply to use the ones that conveniently produce standard charts and graphs.

[12.4]   Statistics, Descriptive and Inferential

We're trying to learn about statistics, but so far, we have only learned to determine one type of average, the mean—and actually just one type of mean, the arithmetic mean. I'd argue that what we have done so far is just calculation; it isn't even really statistical thinking at this point.

I described statistics as *a systematic, mathematically principled methodology for analyzing quantitative data*, but so far we have, at best, just grabbed and used one particular method. Because we know nothing about our data, we don't even know what type of statistics we might be doing.

There are two main kinds of statistics: descriptive and inferential.

In *descriptive* statistics, our set of data is complete and represents the entire *population* we care about. This population doesn't have to consist of people, necessarily. If, for example, someone says to us, "there are six coins in my pocket: two quarters, one dime, and three pennies," then we have essentially *full information* about the coins in that pocket, not just a sample, and then we can begin to discuss that data. Of course, we might decide that the owner of that pocket is more likely to be from the Unites States than from Europe. If the information we have is "there is only one piece of string in my pocket, 1.4 inches long," we might not be able to do much statistical work with this. We might be able to guess that the owner of the pocket is a very unlucky hobbit, I suppose.

For the most part, though, *descriptive* statistics isn't about making these sorts of guesses, however amusing they may be, but about providing exact quantitative results: The person in the first example can pay in exact change for purchases of 63 cents, 62 cents, 61 cents, 60 cents, 53 cents, and so on. The mean value of a coin held by that person is 10.5 cents. Such analysis just tells us something about that pocket and those coins; it doesn't particularly allow us to infer anything else. A true census, involving a head count, will get information from everyone, allowing for descriptive statistics work to be done.

In *inferential* statistics, we have some quantitative information about a *sample* of a population. Polling organizations survey a portion of voters, not every voter; they use this to infer how an election will turn out. Medical tests are seldom done on everyone in the world who has a particular disease or disorder; they are done on a particular group. Inferential statistics work involves extrapolating from the sample (for which we have certain quantitative data) to a larger population.

If we want to consider the ages at which people have been awarded the Nobel Prize in Literature, it would make sense to consider our investigation a descriptive one. Our information should be complete: we know what we know about the entire population; we don't just have a sample.

On the other hand, if we wanted to learn what we can about correspondence (written correspondence, via letters) in a particular country in, say, 1850, or about manuscripts of music from the fourteenth century, it would make more sense to frame our statistical investigation as an inferential one. We don't have all of those letters, and those manuscripts don't all survive. In both cases, we have samples, and we will want to use those samples to learn something about the broader, entire universe of (in this case) inscription.

We've learned to take the average (the arithmetic mean) of ten numbers, using two different programming languages. And we've learned to draw a chart representing those values and that mean in Processing. But I claim that we aren't doing statistics yet because we don't even know whether our data is telling us everything about a population (and thus our activity is descriptive) or is a sample (and thus our activity is inferential). Furthermore, and related to our not knowing anything about the data, we don't know that this particular average we're using is the best one.

[12.5]   The Centers and Spread of a Distribution

Besides the commonly used mean, there are two other fairly well-known averages, two ways to find the center of a distribution. This will all be (or has been) on the test, by the way; the GRE covers these three types of averages and differences between them. And in addition to finding one or more averages, or centers of a distribution, the other very basic way of characterizing a distribution is to determine how spread out or concentrated it is—for instance, using the variance or standard distribution.

[Exercise 12-1]   Median
The *median* is the "middle point," so that if you sort, for instance, 15 data points from smallest to largest, point number 8 will be your median. If you have an even number of

points, there are two "middle points," and the median is these added together divided by two. In [1, 2, 3, 4, 5, 5, 5], the median is 4. If those numbers are jumbled into a different order, it is of course still 4. Averages remain the same whatever the order of the elements; they are properties of unordered distributions, not of sequences.

We haven't yet said much about probability, which is essential to statistics, in this quick discussion of statistical principles and visualization. It's important at this point to at least say what a *distribution* is because it was mentioned a few times in the previous paragraphs. It is the set of possible outcomes weighted by their probability. So, for flipping a fair coin, the distribution is heads (50 percent) and tails (50 percent). In considering sets of data as distributions, each outcome (each element in the set) is equally likely, but particular elements may recur. So in [1, 2, 3, 4, 5, 5, 5], seen as a distribution, 5 is three times as likely an outcome as either 1, 2, 3, or 4. Although hardly a complete idea of distributions—and this quick discussion covers only those that are discrete, rather than ones that might represent very fine-grained measurements, and have arbitrary numbers after the decimal point—this is at least a starting point.

Understanding this, write median(), a function that accepts a list or array of numbers and returns the median. To understand how different programming languages correspond, and how they don't, write this function twice. Write it in Python, beginning with the line def median(data):. Then write it in Processing. In this programming language, float median(float[] data){ will be your first line. How exactly you implement this, in each case, is up to you. You should first develop a function that works with lists or arrays that have an odd number of elements. Then, after this works, revise your function to work whether your number of elements are even or odd. Use Jupyter Notebook for the Python version. As you work in Processing, save several steps of progress in addition to the final function.

For this exercise, write *your own* median() function. Don't use an existing one.

[Exercise 12-2]   Mode

The mode is the most common, or frequently appearing, value in a distribution. For instance, in the previously mentioned set of values [1, 2, 3, 4, 5, 5, 5], the mode is 5. A set of data can have more than one mode; if it has two, it is bimodal, and if it has any number more than one, it is multimodal. For example, the set [10, 2, 8, 4, 0] has five unique values, so (although this may not be very useful to note) it is multimodal and has five modes.

Write mode(), a function that accepts a list or array of numbers and returns the mode. Write it, once again, in both Python and Processing. How you implement it is up to you. It will probably help if you first develop a function that returns a single value, determining *a* mode of the data. In other words, it gives the right answer for a data set with one mode, and it gives one of the modes for multimodal data. Then, after this works, revise your function to work whether you have one mode or many. In revising it, you'll need to consider what *type* of value to return. A single number by itself will no longer work for this second step. Save your progress in addition to the final function, and at the very least save these two steps of development.

For this exercise, you are asked to write *your own* mode() function, not to use an existing one.

[Exercise 12-3]   Variance and Standard Deviation

The other simple way to characterize a distribution, besides finding a center or average of some sort, is by seeing how dispersed or spread out the data points are. For instance, [5, 5, 5, 15, 15, 15] has a mean of 10, and [10, 10, 10, 10, 10, 10] does, too, but these are clearly rather different distributions. In this particular case, the mode would be informative about that difference, but we will not always be so lucky. We'd like to have a general way to describe not just the center of a distribution, but how data are distributed about the center. We can use a statistic called the *variance*. Once variance has been explained, another measure of spread or dispersion will be easy to understand: the *standard deviation,* which is just the square root of the variance.

The variance is the average (and specifically, the mean!) of the squared differences from the mean.

Consider [1, 1, 7]. The mean is $(1 + 1 + 7) / 3 = 3$. Now we need to go through each point and find how far it is from the mean, square that, and average all of those values.

So, 1 is 2 away from 3, and 2^2 is 4. The second value is the same as the first. And 7 is 4 away from 2, and 4^2 is 16. Then we determine the mean of those squared distances: $(4 + 4 + 16) / 3 = 8$. That's our variance. By the way, I said how "far away" each point is as if I were taking a measuring tape and figuring it out that way. But you can just subtract: $1 - 3$ is -2, and -2^2 is 4; $7 - 2$ is 4, and 4^2 is 16. It doesn't matter whether the distances from the mean are positive or negative because they're going to get squared and the final value will end up the same in either case, as the result of squaring $-x$ is the same as the result of squaring x.

The process is similar for any set of data: determine the mean, go through each number, subtracting it from the mean and squaring the result, and then take the mean of this new sequence.

Write variance(), a function that accepts a list or array of numbers and returns the variance. After this, write the very short function stdev() to determine the standard deviation. It should call `variance()`, of course. In Python, exponentiation, as in 2^2, is done with the ** operator; to square two, just type `2 ** 2`. Taking the square root is the same as raising a number to the power of ($\frac{1}{2}$) or 0.5, so the square root of 2 is found with `2 ** 0.5`. In Processing, look up the function `pow()` in the Processing reference to see how to accomplish the same thing.

[12.6] The Meaning of the Mean and Other Averages

Having implemented functions to compute these three averages (mean, median, and mode), let's consider why one would want to determine any of them.

The idea of any of these is to determine a sort of *center* to the data points being considered. This can be used to divide the data into greater-than-average and less-than-average, or close-to-the-average and far-from-the-average, or just to generally characterize the data. If you know that the average stay in your hotel is one week, you won't plan your whole hotel operations assuming that the average stay will be one night or one month.

The mode, the most frequent value, is particularly useful with *discrete* data, as when counting things. The 2000 census determined that the average number of children (people under age 18), per family, is 0.9; that is, the average American family has 0.9 children. (This particular average, as you might guess, is the mean.) This number is useful; if you compare by country, or by state, you can see where more or fewer children per family are being raised. Still, obviously this does not mean that it is typical to have nine-tenths of a child. If we wanted to know the most typical number of children to have, we would use the mode. If we wanted to know the number of children that the statistical "middle American" had, we would determine the median. In both of these other cases, we would get a whole number.

Another consideration is that the mean is very sensitive to individual data points. If one of them is severely in error, the mean could be severely affected. However, such a problem would not affect the median or mode. Now, if one has carefully gathered each data point and does not believe the extremes to be in error, perhaps it's actually a virtue that the mean is sensitive to each one of them.

The mode is not very sensible for a continuous measurement. For one thing, changing the accuracy of your measurement can radically alter the mode. [`1.0, 1.0, 5.1, 5.3, 5.4`] has a mode of 1.0, but if we lost one significant digit, we would have [`1, 1, 5, 5, 5`] and the mode would be 5. Something like this can happen at each step as significant digits drop off of data.

One last note about these averages: they can work together to give you more information, concisely expressed, about a distribution. If the mean is less than the median (and thus, on a chart such as a histogram, further to the left of it), there will be more mass on the right and the distribution can be described as *skewed to the left* because this is the side where it will be more spread out. (In a histogram, unlike in the simple plot we developed, there is an indication of how many occurrences there are of point 1, point 2, point 3, and so on, in order.) The opposite is true if the median is less than the mean; in that case, the distribution is skewed to the right.

Our explorations here will not provide a full course—even an introductory course—in statistics. But I hope that this brief discussion at least helps to explain that these three averages are not simply different blades of Swiss Army knife to be used interchangeably for prodding and prying. They are different methods that should, however they are used, fit into a methodology and be used systematically, with a sensitivity for discrete and continuous data and for the underlying reason that one is seeking an average.

[12.7]   Gathering and Preparing Data

Because this book is about programming, our foray into statistics began by writing some code. But the real starting point for statistical investigation is gathering, preparing, and intelligently using data. We might examine some data and find shocking results about the percentage of people who suffer harmful accidents. If we later learned that this data is not from an unbiased sample of the population, but from patients presenting in emergency rooms, we might be much less shocked.

Instead of considering data collected from a human population, let's return to an issue of editorially prepared data. We've discussed Project Gutenberg e-texts and some issues with using these as data. Legal information has been added to these texts; it's typical that new editions, print or otherwise, will include such material, along with other front matter. We can choose different formats, and some of them will model different aspects of the particular text they are based on. HTML versions can include italics, whereas plain-text versions do not. Typographically, you may find that the inch mark (") is used instead of the opening and closing double-quotes (" ") that actually appear in the original text, and similarly for single quotes, although many plain-text versions have now been updated with typographical quotation marks. Lines are not broken in the same place where they are in the original text, and all sorts of paratexts and material qualities of the text (running heads, page numbers, the particular page-by-page layout) are missing.

These qualities are missing, of course, because this is an edition and a model of a particular earlier text, not a facsimile reproduction of it. (Plenty of those can be found online, too.) Just as programmers choose what to model and what to intentionally overlook as they develop abstractions, editors preparing a new edition of a book must deal with this issue.

Project Gutenberg texts are not simply wrong, whichever types of quotation marks are used. There are editorial decisions made and editorial practices followed in putting together these digital books, each of which can be better or worse from different standpoints. In using these books as data, it's important to understand the abstractions that these e-books are and what can and cannot be learned from them. This may mean choosing the HTML version of the e-text, if italicization is important, or choosing images of pages, if visual qualities of the book are important. You might choose to use an edition prepared in accordance with the Text Encoding Initiative, the heavy-duty markup method for humanistic study. Or your research might bring together more than one type of edition. Finally, if one wishes to study aspects of a work that simply are not encoded in available modern editions, it may be time to gather new data yourself from primary artifacts—data that you should share so that other researchers can critique and build upon your work.

The preparation of data involves gathering it or acquiring data that someone else has gathered, cleaning the data to remove errors that do not reflect the underlying phenomenon, and making intelligent use of the data as analysis of it begins. There are numerous books on data gathering and cleaning; the ways you will engage with these depend on whether you are gathering born-digital data online, using sensors to acquire data, surveying people, or looking to digitized texts and images that have been prepared with different sorts of editorial and curatorial care. But in any inquiry, there is plenty to be done before the statistical analysis and plotting.

[12.8]　Probability and Generating Numbers

You may have noticed that courses introducing statistics are typically called "Introduction to Statistics and Probability." This is not just a convenient two-pack. There is a deep, underlying relationship between the two, and one can begin to get a feel for it by considering that statistics works on samples of data. Probabilistic processes can be seen as generating such data. A major point of statistics is to not just declare things about this data, but to try to determine what process generated it.

For instance, let's say you are given a long sequence beginning with:

tails, heads, tails, heads, heads, heads, tails, tails, tails, heads . . .

Let's say you have another long sequence beginning this way:

    female, female, male, female, female, male, male, male, female, female . . .

Finally, let's say you have another long sequence beginning:

    right, right, right, right, right, right, left, right, right, right . . .

We don't need to know where these sequences came from to discuss them, but to be concrete about things, consider that the first is a record of actually flipping a coin and recording whether the coin comes up heads or tails. The second indicates whether singleton children (i.e., children who are not twins, triplets, etc.) born in a particular hospital were classified as male or female at birth. The third and final sequence indicates whether people in a particular city are more left-handed or more right-handed.

We could critique these examples as only offering two choices in all of these cases. A coin might land facing one way, be knocked off the table, and land facing the other way. Is that two flips or one? Or is the whole thing invalid? Intersex children are born. Some people are ambidextrous or switch between dominant hands depending upon what they are doing. The data we have, like all data, is an abstraction involving choices, not a perfect image of reality. While remaining aware of this, let's consider that, given the choice of abstraction that has been made, the particular elements in the sequence are as accurate as they can be.

What we could notice is that the first two sequences, considered as samples from a probability distribution, are very similar—in fact, basically the same. The process underlying them can be described as follows: *Choose each time between two equally likely options.* To describe it more generally, this is the *uniform distribution* over two discrete values. You can have a uniform distribution over three values, too: choosing with equal probability between rock, paper, and scissors is usually a very good idea. In fact, a uniform distribution can be over any number of discrete options or over some continuous interval.

Now that we have noticed that the first two sequences not only look similar, but, in fact, seem to be generated by the same distribution, we can also note that the third sequence is not generated by this distribution. Left-handedness and right-handedness are not equally common in the world at large or in the populations of different nations or cities.

There is another reason the third distribution is different. In the first two, each value is independent of each other value. Assuming that some coin flip didn't physically damage the coin and imbalance it (which seems a fair assumption), whether a coin comes up heads or tails on one flip has nothing to do with the other flips. Whether a boy or a girl was just born has nothing to do with the next birth at the hospital.

Well, almost nothing. Recall that I specified *singleton* births. If a woman gives birth to several children at once, these children in some cases will be genetically identical. And in that very specific case, there is a strong connection between several children born in order at that hospital. The children will be of the same sex. However, the data that we're discussing do not contain multiple births.

For coin flips and births, it doesn't matter how I sort the data based on underlying features of the data. That is, if I decided to put every even-numbered coin flip first, and then every odd-numbered coin flip, I would still see results consistent with the uniform distribution. If I ordered births according to time of day, I would still see results consistent with the uniform distribution. Obviously, if I put all the heads first or all the female children first, the sequence that results doesn't look much like it was drawn from the uniform distribution, but that's because this involves sorting the data *after* seeing what values appear each time. There's no way I can look at the underlying description of each point and put them into an order like that, because whether I get heads or tails, or male or female, doesn't depend on anything except a random draw from between those two equally likely options.

But for handedness, there is a different situation. Handedness is genetically linked. So, if I were to sort the data by last name, for instance, I would expect to see more clumps of left-handed people that I would if the data were arranged randomly. Also, left-handedness is more common in men, so if the data were organized with all the women first and then all the men, the two halves would differ. This distribution, therefore, is different from the uniform distribution in more than one way. Not only is left-handedness much less often observed than right-handedness; the handedness of different people is not independent.

This is an important connection to make, between probabilistic processes and the generation of data (on the one hand) and statistical analysis to determine what underlying process generated data (on the other). It means, among other things, that the random text generator you worked to modify in chapter 3 has something to do with statistics!

[Free Project 12-1]   Reweight Your Text Generator

Do this project at least two (2) times. Return to your modified text generator from free project 3-1, Modify a Simple Text Machine. Now, as a first step, make extensive changes to the distribution of text it uses—without changing the set of texts. You can accomplish this by simply repeating strings, so that some are selected more often than others. For instance, if you are choosing between the two elements in the list `['Awesome!', 'Oops!']`, each will be chosen equally often. If you change this to ['Awesome!', 'Awesome!', 'Awesome!', 'Awesome!', 'Awesome!', 'Awesome!', 'Awesome!', 'Oops!'], there are still only two possible texts, but the first response becomes seven times more likely. If you changed the list to ['Awesome!', 'Oops!', 'Oops!', 'Oops!', 'Oops!', 'Oops!', 'Oops!', 'Oops!'], you would weight the distribution in a different way. This might make the difference between representing a highly skilled videogame player and one who is a klutz.

The main goal is to find interesting new weightings, but you should also see if you can rewrite your program and find ways to avoid duplicating strings. When you have the same string repeated seven (or more) times, it's fairly easy for an inadvertent keystroke to introduce a typo to one of those copies. Also, you'll end up copying, pasting, and counting a lot of text to determine the weight of a particular outcome, which could be expressed concisely if the program were written more clearly.

After you explore reweightings of the existing (unedited) texts that you have written, continue by revising those texts: change the strings to other strings, in whatever way you would like. Develop a new generator through this process of modification that incorporates new texts and some nonuniform weightings. Try to let the qualities of the distribution, not just the words you have selected, be important to the eventual outcome.

[12.9]   Correlations and Causality

To close this chapter, some discussion (without any further equations or programming) of where statistics and visualization can take you—where it no doubt will take you, if you go further with it.

There's not a huge amount of benefit to just plotting data and averages, as we have done here. The benefit I hope to offer is allowing you to understand some statistical principles, however simple, and to see how they work together and can connect with programming and visual design. But plotting averages isn't going to do much intellectual heavy lifting.

If you have lots of data, you really want to find ways in which it's correlated. For instance, you might have some news feed data and some message/status/posting text from people, with information about the time of posting in both cases. We might expect that people will write to each other about news events, but is that the only correlation? Does affect or self-declared mood vary in a way that is correlated to news writing? Do people use more question marks or make more all-caps statements at some times? Or are there some news stories that follow (in time, and also in causality) online discussion—either because they are explicit about online discussion or in more subtle ways?

Visualization, with statistics related to one data source on the x-axis and those related to the others on the y-axis, can provide a nice way to see where hot spots of correlation are. You could also read through a grid of numbers; it's a convenience to be able to visualize them, though.

Although this book does not cover how to determine correlations, there are different ways that are not too conceptually or mathematically heavy. As with the different averages, the process of finding correlations should involve assessing these different measures and understanding the principles upon which they are based.

Perhaps the most hot-button issue involving correlation that of its relationship to causality. The reason statisticians look for correlations is, certainly, because they are interested in learning about causality. However, seeing correlation may not indicate a causal connection, and it never reveals, by itself, the mechanism of causality.

Let's say we have A and B that look like they are reasonably correlated. Here are six reasons they might appear this way:

1. **A causes B,** maybe through one or more intermediate steps. For instance, data about a sample of a population shows that being recently convicted of a felony is correlated with keeping a fixed residence for a while. The people we are considering are mostly in prison. Being convicted led to their being in prison, and being in prison means staying put.

2. **B causes A,** similarly. It's not always obvious in which direction causality goes. For instance, a researcher might observe that illegal drug use is correlated with other criminal behavior. Someone's first guess might be that doing drugs *causes* people to commit crimes. But perhaps people who start off committing other crimes begin to use illegal drugs essentially because of peer pressure, because such drugs are easily available to criminals, or even because doing drugs is necessary to establish the trust of other criminals. Even if there is a causal link between A and B, this doesn't establish the direction of the causality.

3. **There is some common cause for both A and B.** An example might be sales of window air conditioner units and deaths from heatstroke. This correlation might seem odd, because people cooling themselves with their new air conditioners would seem to be less susceptible to heatstroke. It's not the same people making the purchases and dying from heatstroke, for the most part, however. What's happening here is that an underlying factor, temperature, varies seasonally and due to extremes such as a heat wave. Both A and B increase as the temperature increases and decrease as it decreases, although different parts of the population are involved in A and B.

4. **A and B reinforce one another.** This case is similar to case 3 but might be distinguished. Having higher income might lead to better health (in the United States) because one can afford better health care or perhaps higher-quality food and improved maintenance of wellness. But poor health could also keep people from working and earning money and thus cause financial problems.

5. **A and B are measurements of the same thing.** If you look at a temperature in Fahrenheit and the same temperature in Celsius, you'll find that they appear strongly correlated. Perhaps perfectly correlated. That's because they are the same measurement with different units. A and B could be less than perfectly correlated in this case due to measurement or recording error. This is perhaps the limit case of case 3 but certainly is an extreme situation.

6. **There is no underlying reason for the appearance of correlation.** In the previous cases, there is some sort of causal link or links. But the reason for correlation—or, to be precise, the reason for the appearance of correlation—might be no reason at all. There's a strong correlation between the consumption of chocolate in a country and how many Nobel Prize winners that country has. Between 1997 and 2007 (at least), autism diagnoses in the United States increased at almost exactly the same rate as organic food sales. Perhaps most famously, the decrease in incidents of (true, nautical) piracy almost exactly match the increase in average global temperatures, a striking case of *anticorrelation* between piracy and global warming. Statistically, the data seem connected, but consider: if we start with A and we then search through a *million* different possibilities and find one of them, B, that is correlated with A, the fact that we found this one correlation may simply be meaningless; it might just be due to chance. With a large enough quantity of randomly generated data sets, or data sets found in the world, we are sure to get some sort of appearance of correlation somewhere. But that doesn't really mean that there's any relationship at all. This is important to keep in mind if you are searching among a really vast space of factors and find a few correlations that don't seem to make sense to you.

Correlation is a signal—often a good signal—that there's some sort of causal link there and that further investigation should show what it is. But it doesn't establish that link, or what kind there is, by itself. It's the beginning of inquiry rather than a resounding, final conclusion.

[12.10]   More with Statistics, Visualization, and Processing

If you plan to work with statistics, some introductory study of the field would probably be very valuable. This could come in the form of a classroom course, an online course, or reading and working through an introductory book on statistics, such as David A. Freedman's *Statistical Models: Theory and Practice.*

To work with visualization in Processing further, see Ben Fry's *Visualizing Data: Exploring and Explaining Data with the Processing Environment.*

There are also several other good books on Processing. Unlike many books focused on specific programming languages, they take an art and design perspective. This makes for a pleasant journey through the language as these books explain and demonstrate Processing in relation to programming principles and elements of visual design. You have enough experience with Processing at this point to explore on your own, but if you feel that you want to follow along in a book through more of Processing, consider *Getting Started with Processing* by Casey Reas and Ben Fry; *Learning Processing: A Beginner's Guide to Programming Images, Animation, and Interaction* by Daniel Shiffman; and *The Nature of Code: Simulating Natural Systems with Processing* by Daniel Shiffman. The sort of introduction that is provided by the Processing reference and examples is offered, alongside essays and interviews, in *Processing: A Programming Handbook for Visual Designers and Artists* by Casey Reas and Ben Fry.

Processing has a lot of graphics capabilities built in, so you can do a great deal with it without locating new libraries for it. However, to get further into statistics work, you will probably want to use the Papaya library:

adilapapaya.com/papayastatistics/

In Python, our Anaconda distribution is purpose-built for managing data, doing statistical analysis, and creating visualizations. For instance, it includes NumPy (Numeric Python, supporting matrix operations), SciPy (Scientific Python, with plenty of statistical functions), and matplotlib for plotting.

[Free Project 12-2]   An End-to-End Statistical Exploration

Prepare some data—for instance, text files or images. You have, of course, been working with data of these sorts already. For this exercise, it is preferable to use *born-digital* data rather than older data that has been digitized because you have control over how to prepare such data rather than having it extensively prepared for you in a certain way. However, if your interests are in earlier times and there is relevant digitized data that you wish to explore, you can use this after you investigate how it was digitized and prepared. If you are using data that you gathered, clean and prepare it, making sure that this doesn't lead you to lose file creation time information or other important metadata.

Then, write a program in Python or Processing that yields statistics of this data. If you wish, visualize the original data or statistics of it; visualization is a means, not an end, so this isn't a firm requirement. Whether by looking at numbers or a visual display, see what you can identify that deserves further investigation. (*This* is the end, the ideal outcome, of an exploration.) See what values seem surprising. Is there a great deal of variance where you would expect little? Do some values that you expect to be correlated seem to have no connection to one another, or is there *anticorrelation* (one value rising as another falls) when you expected the opposite? Feel free at this stage to determine correlations by using tools such as the stats module of SciPy, by inspection of numbers that result, or by inspection of plots of those numbers.

The goal of this exercise is not to use statistics to come to particular conclusions, but to use statistics to explore. A successful result would be noticing one or more surprising, unanticipated aspects of the matter you are investigating. So, you do not need to write a paper about your topic and develop conclusions to complete this free project. The end being sought here is not that final. You should just identify some new, unusual qualities (and quantities) in the data.

If after finding something unusual, you do continue and investigate further, you may discover some surprising underlying causes. You could end up writing about this phenomenon without even mentioning the statistical analysis that led you to focus on this particular aspect, just as a poem developed with computational assistance does not have to be published along with the code that helped the author write it.

[12.11]   Essential Concepts

[Concept 12-1]   Programming Fundamentals Span Languages
There certainly are differences between Python and Processing, but the core aspects of programming—including iteration, functions, and types—are essentially the same, even though they are expressed differently in the two languages. Once you learn the basics of programming (at least, programming in an imperative idiom, which is our focus in this book), you can apply what you know by figuring out the differences in syntax. It's not necessary to learn new fundamentals. Be sure you can deal with iteration, functions, types, and the conditional in Processing as well as in Python.

[Concept 12-2]   Different Averages Have Different Meanings
Understand not only how to compute the mean, median, and mode and how to determine the spread of a distribution via variance and standard deviation, but also what these simple statistics express about a distribution and in what cases they are meaningful.

[Concept 12-3]   Probability and Statistics: Sides of the Same Coin

Perhaps even better: One of them can be said to govern the coin flip, while the other is a way of analyzing flipped coins and learning about what led to those outcomes. Understand how to use probability to generate "distributional" digital art (textual and otherwise) and how to use at least the basics of statistical analysis to characterize a distribution.

[Concept 12-4]   Visualization Should Be Principled

It's fine to produce an aesthetically pleasing display of data that is nice to look at for its own sake. But visualization can also be connected to statistical principles, allowing the researcher's sight to help distinguish what aspects of data are the most interesting. Understand how the design of a simple visualization can facilitate better-grounded statistical thinking.

[13] Classification

[13.1]

Classification is the process of categorizing data into discrete sets. We have already covered some extremely simplistic cases of this process. In 7.5, Converting a Number to Its Sign, we figured out how a function could classify numbers as positive, negative, or zero. Also in that section, we saw how a classifier for the fictional twofold land of Binaria could categorize Binarian names. Even with a very simple example of this sort, it was possible to see that classification is a culturally loaded process with implications for those who do not wish to fit into the classification scheme.

In this chapter, we'll first discuss how to classify more complex data (books, represented as text files), initially using simple hand-crafted conditions. Then, we will use a classifier that is provided by the TextBlob library to classify texts. Finally, we will look at how some very simple sorts of image classification can be done, too, and provide pointers to more typical ways of classifying images.

If you chose not to install TextBlob initially when you were first getting software set up, you will now need to return to 2.5, Install a Python Library: TextBlob, and follow the instructions there.

[13.2] Verse/Prose Text Classification

Of the many types of text analysis we can conduct, one of the most straightforward involves placing a text in one of a few known categories. The simplest case is when there are two categories, as when we might want to distinguish fiction from nonfiction. Let's consider something that might be even more immediately visible to a human reader just glancing at a text: classification of a text as poetry or prose. Or, to be more specific, as either *verse* (lineated language) or *prose* (in those well-known rectangular paragraphs, particularly visible when the text is justified). We won't ask our system to distinguish prose poems from short stories, but we will see how much effort it takes to distinguish lineated language from language in paragraphs.

As with any computation, it's possible to start figuring out how to implement a verse/prose classifier by asking how a human reader would do this classification by hand, or, more likely, by eye. A quick look at the Project Gutenberg editions of *Pride and Prejudice* and *Leaves of Grass* (obtained back in chapter 9, "Text II: Regular Expressions") might prompt one or more of these ideas:

- Each line in prose paragraphs runs almost to the end (72 characters) until the final line is reached.
- There are blank lines between prose paragraphs.
- There are also blank lines between poems and stanzas.
- Some verse lines are much shorter.
- Verse lines are irregular.
- Poems have titles that aren't indented, while the text of the poem is.
- Groups of prose paragraphs are separated by numbered chapter headings.

Some of these are rather specific to the particular files selected. Whitman's free verse lines are more irregular than is metrical verse (*Paradise Lost*, Shakespeare's plays or sonnets, etc.), even though what is being counted out in meter is the number of feet, not the number of letters. A long poem such as *The Odyssey* might have numbered sections (books) that look like chapter headings instead of having individual poem titles, while some other long prose works (such as short story collections) are not divided into chapters. Blank lines occur in various places, and there is no reason to think they will consistently be more or less frequent in one type of text. But it's reasonable to consider that prose typically runs, in its rectangular way, consistently to the right side of the page, resulting in lines that are long and fairly regular.

Okay, next, load up *Pride and Prejudice* and *Leaves of Grass* in a Jupyter Notebook session that you start in the appropriate directory, where the two books are located. This time, because we care about line length, we'll use the `readlines()` method instead of `read()`. The latter method placed the entire contents of the file in a single string. We could go through this string and look for newlines, but using `readlines()` is easier. It provides a list of strings (with each string a line of the file):

```
source = open('1342-0.txt', encoding='utf-8')

pride = source.readlines()
```

```
source.close()

source = open('1322-0.txt', encoding='utf-8')

leaves = source.readlines()

source.close()↵
```

The extra argument to open(), specifying the encoding, is okay to use but not necessary if you are running GNU/Linux or Mac OS X. Back in 6.4, Iteration (Looping) Abstracts along Sequences, we looked at the following function to determine the mean of a list of numbers:

```
def mean(sequence):

    total = 0

    for element in sequence:

        total = total + element

    return total / len(sequence)
```

This one should be familiar from the previous chapter, too! Let's modify this slightly. We will simply substitute len(line) for element. Instead of accepting a list of *numbers*, this function accepts a list of *strings* (the lines in the file) and determines the length of each one. Enter this version into Jupyter Notebook:

```
def mean_line_length(sequence):

    total = 0

    for line in sequence:

        total = total + len(line)

    return total / len(sequence)↵
```

I did actually made two other changes. Notice that there was no requirement that I change the name element to line. For that matter, the function's name didn't have to change to mean_line_length(). I just thought it would be a bit clearer to make these two changes. The only thing that *has* to change for the function to work as desired is that len() has to be called in the second-to-last line.

Now check the following:

```
mean_line_length(pride)↵

mean_line_length(leaves)↵
```

There's a difference, certainly. It might be even clearer if we didn't count blank lines. Such lines probably aren't going to be informative because both prose and verse texts have them, and seemingly either one could have more or less. To skip them, we'll have to modify mean_line_length(). We can find out when a line stored in the variable line is blank by checking line.strip(), which returns a string that has all leading and trailing whitespace removed. If line.strip() has something in it, we'll continue to process the line. We can use a simple conditional statement, if line.strip():. If there's something in line, line.strip() will be True in this context, but if line has only whitespace or is completely empty, the result will be False.

```
def mean_line_length(sequence):

    total = 0

    nonblank = 0

    for line in sequence:

        if line.strip():

            total = total + len(line)

            nonblank = nonblank + 1

    return total / nonblank
```

If you call mean_line_length() on both books after making this change, you'll see that the average lengths have been slightly pushed apart, by less than a single character, but by more than half a character. So perhaps omitting blank lines doesn't really matter. Whether or not it does—and you'd really want to check some other data to figure that out—what we did here is to try it out, to test our guess empirically.

For this exercise in manually building a classifier, we can find out what value is directly between these two average values.

```
(mean_line_length(pride) + mean_line_length(leaves)) / 2
```

Make note of what this is, at least to two decimal places, and put it into this simple classifier, in the blank:

```
def verse_or_prose(text):

    if mean_line_length(text) < ____:

        return "verse"

    else:

        return "prose"
```

What results is a classifier—in that it does indeed put any Project Gutenberg e-book into one category or the other. This is a *trivial* classifier in that it uses only one feature. You can't have one that is simpler. In addition, the dividing line between verse and prose has been set in the simplest possible way, by inspecting only two data points.

You should check `verse_or_prose(pride)` and `verse_or_prose(leaves)` to make sure that this classifier is wired up correctly for the two examples you looked at. And after you check that and make sure that it's working, load up *Moby-Dick* (be sure to use `readlines()` as you did with the other two books) into the variable moby. Now, for the moment of truth:

```
verse_or_prose(moby) ↵
```

Admittedly, you can't determine much from checking one additional book. To do any real sort of test this classifier, you would need a corpus of books, a set of test data— even if it is a tiny one, at first.

[13.3]   A Miniature Corpus for Training and Testing

To see how a corpus can be used in setting a dividing line and in testing, we'll scale up to what is still a very small-scale case, considering how you would use a corpus of only ten documents in total. The extended exercise in this section is still much more limited than in any practical training and testing situation, but it illustrates the principles of classification at work. It shows how to enlarge your corpus beyond the trivial case of two training documents and one test document. It is an exercise rather than a free project because the number of books, the type of classification, and the features that you use for classification are all specified. The only thing that is "free" about it is that you should select your own books to form your own corpus of documents.

[Exercise 13-1]   Train and Test on Ten Books

Download ten e-books from Project Gutenberg, five of which are prose books (they don't need to be fiction, just prose) and five of which are books of verse (poetry broken into lines). These can otherwise be of your choosing. Be sure they are in a plain text format, not in HTML. You can use *Moby-Dick* and *Pride and Prejudice* as two of your prose books, if you like, and *Leaves of Grass* as one of your poetry books. I've used these texts during some of the times and other texts at other times when I went through this exercise. Whatever you choose, don't select books that mix both types of writing.

Clean your data by removing the Project Gutenberg legal information along with all other sorts of apparatus, all the front matter and end matter: editorial introductions, prefaces, and so on. You want to have just the core text of these ten books when you've finished.

Next, set aside three verse books and three prose books to use as *training* data, or at least what will approximate training data. You will manually set a dividing line based on the properties of this data. The other four books will be used to *test* your classifier, so you can see if what you developed actually worked, at least in these cases. You will not do anything with these books until you have developed a classifier using the training data.

Using your mean_line_length() function, determine the average (arithmetic mean) length of the lines of the three training prose files. Then, average that—you should use your mean() function—into your prose mean. Use mean_line_length() to determine the mean line length of your three training verse files. Then, average that into your verse mean. Midway between the prose mean and the verse mean will be the dividing line your classifier uses when it comes to mean line length.

For this exercise, we'll incorporate one additional feature that we'll arbitrarily give equal weight. In addition to considering the mean line length, the other feature we'll use is the variance in line lengths. Use what you learned about statistics in the previous chapter, and the way mean_line_length() works, to write a function called line_length_variance() that will return the appropriate value for a document—that is, a Project Gutenberg e-book. In the same way you manually found a dividing line between prose and verse means, find a dividing line between prose- and verse-line length variances.

Now, set your prose_verse() classifier so that mean line length is worth 50 percent and line length variance is worth 50 percent as it makes classifications. Is it clear how to do this? It can be done by multiplying each feature by the appropriate amount so that each of the two features gets an equal vote.

This type of two-feature classification is the sort that might be used in a pass/fail university course. Imagine a course in which there is an exam and a paper, each worth 50 percent of the grade, used to distinguish between students who pass and students who fail. The exam and the paper are two different "features" that the instructor is using, with equal weight, to classify students. Of course, if the test grades and paper grades are on different scales, they will need to be normalized. Normalization involves dividing by whatever the maximum possible value is. If the maximum exam score is 25, while the maximum paper score is 100, normalizing these two grades properly will involve dividing by those values.

After you have developed a classifier, the first thing to check is to see if all your training data is correctly classified. It may not be. In some cases, no line will divide the two categories based on the features selected. If the classifier doesn't work perfectly with the training data, it is very unlikely to generalize well to other data.

After checking the training data, try the test data with prose_verse() and see what results you get. Remember that the point of this exercise is to go through all the stages of developing a very bare-bones classifier, with two manually determined features that are manually set, and then testing the classifier on data that you have held out. If your classifier doesn't work on certain examples, consider why the two features we used may not be good ones to distinguish prose from verse when it comes to those particular documents.

[13.4] Building Up the Classification Concept

A classifier categorizes; in the case of a two-way or binary classifier, it divides a set of data in two. Usually this is not done using a single feature, as was done initially, or two features, as we did in the extended exercise in the last section. Rather than dividing a line at a single point, a classifier can be seen as doing something more general: Dividing a space (of any dimension) using some surface (of one dimension less than the overall space). Strictly speaking, and to make this sound as cool as possible, a binary classifier divides data using a hypersurface. In this most recent exercise, we had just two dimensions of data—the mean line length and the line length variance—and the classifier you developed, manually, will divide this data using a line.

Implicitly or explicitly, training a classifier involves finding one or (almost always) more *features* that will be used to distinguish the data. In this case of the first verse/prose classifier, only one feature was used, the mean line length. You extended this in the exercise to two features, the mean of the line lengths and the variance of the line lengths. A classifier that works on texts, such as emails, will often have many, many more features; as we'll see, a fairly simple classifier will have one per word in the overall vocabulary.

It's important that some *test data* be held out when constructing a classifier, as was done (trivially) with *Moby-Dick* at first and as was done with the two verse and two prose files in the exercise. You can't get any idea of how well your classifier works unless you try it on test data of this sort. If you do keep some test data set aside, you can try different features or otherwise tune your classifier to produce better results. You need to always hold out some test data, unless you truly don't care about how good of a job your classifier is doing. And if that were the case, why would you bother to train a classifier at all?

You probably don't need a computer system to tell you whether *Moby-Dick* is verse or prose, but it may nevertheless be interesting to test the hypothesis that average line

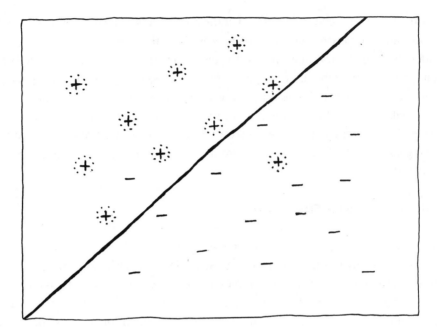

[Figure 13-1]
Training a binary classifier defines a boundary between positive (+) and negative examples (-). The simplest boundary for one-dimensional data is a point; for two-dimensional data, it's a line such as the thick one here. A classifier could however be trained to learn other sorts of boundaries, such as the dotted circles. If it is trained this way, it will score perfectly on all the training data! But it's almost certainly inferior, a victim of overfitting, and won't generalize well to new data. Using held-out test data guards against this sort of problem.

length is sufficient for distinguishing these two categories of text. To develop a very simple classifier, we only needed to use code discussed before (to determine the mean); no sophisticated linguistic systems were employed. In the previous exercise, we went beyond that to incorporate both the mean line length and the variance in line length. Now that we have seen what a classifier is, we'll move along to look at how far more difficult distinctions—ones that might take human readers a considerable amount of time—can be made automatically.

[13.5]   Text Classification and Sentiment Analysis

Let's now move to a different, difficult problem, one that seems to go significantly beyond the lexical level and even the structures of syntax. Texts signal to us, in various ways, what the attitude (or as a first step, the *polarity*) of the writer is toward the subject matter being discussed. We understand a movie review to be negative or positive not only by virtue of how many stars are assigned by the reviewer, but also because of what the reviewer writes. In analyzing language, and particularly contemporary language found online, it could be useful to systematically determine the *sentiment* of texts.

It would seem that counting words, or adjectives, and even knowing the parts of speech of each word is not going to be a huge help by itself. Of course, it's an empirical question. It could be that on a certain message board, blog, or site, the people responding negatively are always snippy and write very short comments. People who are positive always write longer, effusive comments. And so, if we trained a system on this data, marking up which comments are positive and which negative, we might find that there's a very good dividing line at, for instance, 14.5 words and that this single feature—the length of a text—suffices.

Such a finding would reveal the value of *feature engineering*, which is selecting what features will be examined when training a classifier or other machine-learning system. This could be the number of words, number of exclamation points, number of prepositional phrases per sentence, and so on. A feature can, of course, be mean line length, as discussed in the verse/prose example. In addition, a feature can be the overall length of the document, in words or characters, which might distinguish some negative texts from positive ones. This feature, however, is unlikely to generalize very well. In a newspaper in which all movie reviews are about the same length, for instance, it won't be possible to distinguish negative and positive reviews based on length.

If we were interested in building a sentiment analysis system from scratch—a highly polished and truly useful one—we will certainly need to think about feature engineering. If we are improving such a classifier for use with specific data, or developing a new classifier for our own purposes, we similarly will need to keep features in mind.

[13.6] Training on Positive Words and Negative Words

As a starting point, we can simply use a set of words as features so that our program considers how word occurrences differ between positive and negative texts. (Our view of words, strictly speaking, is that of linguistic *types*, as discussed in detail in 15.2, Words and Sentences.) In doing this, we make a common assumption in language analysis—one that has been made in topic modeling, for instance, as well. We assume that a text is an unordered set of words, a "bag of words." The sequence in which the words appear will be ignored.

I'll explain before too long why this approach is limited and cannot capture all types of variation in sentiment. The vocabulary found in enthusiastic, upbeat utterances or texts does indeed differ from that produced by haters, however. And in some cases, reference to that vocabulary may even be enough—just as sometimes, one can happen to successfully split a text into sentences with a simple split('.').

Imagine a system that just looks at words and considers texts as bags of words. *Awesome* will probably have a strong positive score and *icky* a strong negative score, and, although many other words might be neutral, in the end we can approximate the sentiment of a text from the sentiment of its words.

Now, in distinguishing shorter and longer lines, we used a classification function that was incredibly simple. But in creating classifiers more generally, the simplest practical method involves constructing a naïve Bayes classifier. This is the same type of classifier often used in email systems to distinguish spam messages from those that aren't spam (which, in the business, are called *ham*).

The naïve Bayes classifier, by name, refers to the framework of Bayesian probability. It is a good, simple classifier that is actually used in practical applications. We'll consider the simplest variation of this model, a binary classifier. But we will use more than one feature. We'll use a large number—one per word. So *awesome* will be a feature, and so will *icky*.

The extreme and very useful assumption made by our type of classifier is that each feature is independent (conditionally independent, to be precise) from all the others. This means, specifically, that we assume the occurrence of the word *doornail* does not make it more or less likely that we will see the word *dead* in the document, seeing *laptop* does not increase our chances of seeing *computer*, and seeing *quarterback* does not indicate anything about whether we should see *touchdown*. Obviously, words are *not* conditionally independent of one another. However, making the assumption of conditional independence turns out to work very well in many cases.

There is a good and fairly short TextBlob tutorial on how to create your own classifier, one for sentiment. It is called "Tutorial: Building a Text Classification System" and is located here:

textblob.readthedocs.org/en/dev/classifiers.html

The tutorial shows how to create training and test data and how to create a new naïve Bayes classifier trained on the training data.

To understand how to use TextBlob, go and read this brief tutorial. Follow the instructions to import NaiveBayesClassifier. You'll then create a new classifier called cl using training data, as shown in the next line of the tutorial. If you find you are missing an NLTK resource that you need, just follow the instructions in your Jupyter Notebook to obtain that resource.

You can use the training and test data provided in the tutorial, but here is some of my own training data (or test data, if you like) for you to also try out:

```
sentiments = [

    ('Wittgenstein wrote one of the greatest philosophical works ever, an
    incredible contribution.', 'pos'),

    ('The Oulipo is a radical, pioneering group that has shaped literary
    history.', 'pos'),

    ('What an awesome sunset.', 'pos'),

    ('I love it!', 'pos'),

    ('Very good plan.', 'pos'),

    ('The final season of Game of Thrones made my eyes bleed.', 'neg'),

    ('Movies based on DC comic books are extremely tiresome.', 'neg'),

    ('That is a horrible idea.', 'neg'),

    ('I hate that sort of thing.', 'neg'),

    ('You lack imagination.', 'neg')]↵
```

There are several words that suggest a positive sentiment (*greatest, incredible, pioneering, awesome, good,* as well as possibly *contribution* and *shaped*) and some that intuitively suggest a negative sentiment (*tiresome, horrible, hate*). Some of the negativity, such as the figuration for crying in the sentence about *Game of Thrones,* might be very difficult to discern at the word level. (This particular sentence, which I wrote especially for the second edition, also reveals how long it takes to get a textbook revision through the publication process.) Perhaps more disturbingly, if we continued along the same lines and amassed numerous examples like these, we might train our system to find all references to *high cultural* matters to be positive and all references to *popular culture* to be negative. That might be how particular cultural commentators feel, but again, it would not allow us to correctly classify texts in general based on their sentiment.

Let's not speculate. The tutorial says we can use `cl.show_informative_features(n)` to show us the n most informative features in that classifier. If you type in my training data, train a classifier on it, and look at 10 or 20 of the features that have most weight, you may be surprised! There are only three words out in front; every other word, if informative at all, is equally informative. It's unlikely you would guess that any of these words would particularly indicate a positive or negative sentiment. And perhaps they don't, or perhaps this small amount of training data is way too little.

After you have trained a classifier, you can use its `classify()` method to give you a simple categorization (`'pos'` or `'neg'`) for any text tested. You can also see what probability the classifier is assigning, for a particular text, to each of the two categories.

Try adding to the training data with the `update()` method, check the accuracy on the test data, and try classifying sentences of your own.

At the end of this TextBlob tutorial on classifiers, the Feature Extractors section explains that it is possible to expand the default features (simply the occurrence of particular words/types) that are used for classification.

TextBlob provides a rich set of ways to do your own classification. Reasonably good sentiment analysis is *included*, so it's not necessary to build your own classifier to determine the polarity of texts, as we have done. If you wish to, though, you can build your own classifier and even change the features that it uses. In fact, you cannot only go under the hood and work with classifiers in this way. You also can rebuild the engine if you like. The code for the naïve Bayes classifier is written in Python, and, for those of us who have TextBlob installed, it's sitting on our disk drives. We can read it to understand how it works or do even more.

Here's a significant line from the tutorial that gives us a clue about where the source file can be found on our system:

```
from textblob.classifiers import naiveBayesClassifier
```

The first part of this `import` statement means that there will be a file `classifiers.py` somewhere. Specifically, it will be located in a directory called `textblob`. That's what the two parts of `textblob.classifiers` correspond to. To find this file, I used the command-line program `locate`. You can search for it using another method, such as in your GUI. My file (I was using Python 3.7 at the time) was here:

/home/nickm/anaconda3/lib/python3.7/site-packages/textblob/classifiers.py

You could simply open this file in a text editor, as I did. In this file is the code for `naiveBayesClassifier`, beginning (in this version) on line 289. There are just three methods defined there; it looks like the class is mostly a wrapper for the NLTK (Natural Language Toolkit) classifier. But the code for that, too, is right there on disk, in my case at:

/home/nickm/anaconda3/lib/python3.7/site-packages/nltk/classify/naivebayes.py

There's the core of TextBlob's naïve Bayes classifier. At 256 lines, it's not as concise as the code we've been writing, but it's also not beyond understanding. For instance, without getting into fine details, and without asking why this particular code was written, consider lines 112–121:

```
for label in self._labels:

    for (fname, fval) in featureset.items():

        if (label, fname) in self._feature_probdist:

            feature_probs = self._feature_probdist[label, fname]

            logprob[label] += feature_probs.logprob(fval)

        else:

            # nb: This case will never come up if the

            # classifier was created by

            # NaiveBayesClassifier.train().

            logprob[label] += sum_logs([])   # = -INF.
```

You should be able to see that those first two lines are a programming construct you have used: a nested loop. This part of the code does something *for* each label, and within that loop, it does something *for* each item in the feature set. Each item, in this case, is a tuple containing the feature name and the feature value. You learned about this data type in 10.2, A New Data Type: Tuples. After this is another familiar construct, the conditional.

My point is not to explicate this code, or even a tiny part of this code, but to show that if you do peek underneath the abstraction layer, you'll find the same essential sort of programming you've already been doing. More elaborate, yes, but using the same principles.

Looking through the files and reading the methods, their arguments, and the beginning of what each does can be informative in some cases. Of course, the documentation should provide all the information that is needed to use this classifier, but Python modules are made of code that is available to whoever wishes to inspect it.

[13.7] A Thought Experiment about Sentiment and Word Order

Consider the following two texts, the first one consisting of three sentences, the second one of two. Do they convey more of a *positive* or more of a *negative* sentiment?

Windows absolutely rocks. It never sucks. Start me up!

Windows ME absolutely sucks rocks. Never start it up!

Those who are fluent in English and know colloquial, contemporary English would have to agree that the first text is definitely expressing a positive sentiment, while the second text is definitely expressing a negative one. The topic of the two texts may be slightly different; the enthusiast who utters the first one may be speaking about the Windows operating system in general, or maybe Windows 95 (the Rolling Stones song "Start Me Up" was used to promote that version), while the second speaker is specifically slagging a later but still archaic version, the Windows Millennium Edition (Windows ME).

Still, the difference in sentiment is pretty clear. The first text is definitely positive, the second one definitely negative. People wouldn't be confused about this.

Notice, however, that these two sentences have exactly the same words, whether "words" is taken to mean types (number of unique lexemes) or tokens (particular occurrences of types), a distinction explained in detail in 15.2, Words and Sentences. The second text is just a rearrangement by word of the first; it can be seen as a "vocabularyclept" text created from the first. Here's an even simpler vocabularyclept example:

This movie was a great experience, and not disappointing.

This movie was not a great experience, and disappointing.

If you consider, as I do, that these are both plausible examples that are representative of the way language works, rather than completely freak occurrences, it's a real problem for the bag of words assumption being applied to sentiment analysis. You might find that for a certain corpus, this assumption does all right—perhaps about as good as human annotators do. But you can be sure that there are texts (and perhaps significant corpora of texts) where these features won't be enough. The only way to test out how much this difference matters is to train two classifiers, one that uses the bag of words assumption and one that takes word order into account, and see how they do.

One way to take order into account is to add features to the handcrafted sentiment classifier that you have built. The Feature Extractors section, at the end of the tutorial, explains the basics of how to do that. Another is to change to an entirely different model of classifier, then train that type. Instead of focusing on either of those methods, though, let's move on to work with the advanced sentiment classifier that already takes word order into account, one developed and trained by natural language processing researchers and included in NLTK and TextBlob.

[13.8] Using the Included Sentiment System

When starting a new session in which the `TextBlob` class hasn't been imported, be sure to import it:

```
from textblob import TextBlob↵
```

Now, let's check to see what the sentiment of two similar sentences is according to TextBlob's built-in, pretrained sentiment analysis.

```
TextBlob('This movie was a great experience, and not disappointing.').sentiment↵
```

Notice that what the line above does is to create a new TextBlob object using that sentence, which is passed to the TextBlob constructor as a string. Then, the sentiment attribute of this object is returned. As written, however, the TextBlob object is just thrown away. This new TextBlob object isn't assigned to a variable, so there's no way to access it after this line of code runs. One would need to create another TextBlob using the same string in order to access any other methods or attributes.

Also, notice that the output isn't simply 'pos' or 'neg', and indeed isn't simply a positive or negative number. It is an object, and when we see it printed out as a string there are two attributes that are exposed to us. The second one is *subjectivity*. It indicates whether (according to the classifier) the text is subjective or not. It doesn't make much sense to ask if "Mass can be converted into energy" has a positive or negative sentiment because it's about as objective a statement as can be made. Try creating a TextBlob using this sentence and checking the sentiment attribute: you should see that the value for subjectivity that is assigned is zero.

I would suggest that not only word order but also significant amounts of context are necessary for anyone (human or computer) to determine whether some particular sentences are subjective or objective. It sounds to me, for instance, that "His character is chaotic good" makes for a completely objective statement, as much as the one about mass and energy, but I have to admit that it's only objective if one is speaking literally about someone's Dungeons & Dragons character. If the statement is being used to describe someone's personality, it would certainly be at least somewhat subjective. The version of the TextBlob sentiment analysis system that I used determined that this text is 0.6 subjective—with the subjectivity in the range from 0 to 1.

Another difficult-to-handle example is "She is a criminal." If the person being referred to has been convicted of a crime, this is presumably quite an objective statement that describes that person's legal status. If the statement is being made figuratively, perhaps of a CEO who has not literally been convicted of a crime, it would have to be seen as subjective. The subjectivity assigned by TextBlob to this sentence is 0.55.

So, subjectivity is not always simple to determine. Taking context into account may ultimately be the right solution, but something that can also help a great deal is training one's subjectivity classifier on data that is similar to the data being classified. Movie reviews were used to train the classifier included with TextBlob, so it will probably happen that words and phrases pertaining to movies and used in reviews will be better identified as subjective or objective. I note that "He is a star" is a completely objective statement (subjectivity of zero) according to the classifier, which is perhaps justifiable in the movie domain but not in other contexts.

Without trying to investigate exactly why these examples are classified the way they are, it's still worth nothing that the source of training data could have something to do with it. Because movie reviews may be more likely to speak of someone's performance as criminal than they are likely to document people who have literally committed crimes, the subjective sense of calling someone a criminal may dominate. Because someone can objectively be called the star of a film, calling someone a star may be recognized as objective.

So that's a bit about subjectivity and why one would want to detect it. Onward to what we came for, polarity. You should have already checked the sentiment (and therefore the polarity) of this sentence:

```
TextBlob('This movie was a great experience, and not disappointing.').sentiment↵
```

Now, try this one:

```
TextBlob('This movie was not a great experience, and disappointing.').sentiment↵
```

The polarity is different: the first sentence is recognized as positive and the second sentence as negative. From this alone, without looking into documentation or code, you can tell that the classifier is somehow taking word order into account. Whether it's doing that with some entirely different classifier model or is simply looking for words that are negated, there is some method that takes into account more than just pure occurrences of tokens.

The polarity is also returned as a number between -1.0 and 1.0. One interpretation of this is that slightly negative comments can be distinguished from very negative ones. Another interpretation would be that the number is providing a measure of confidence, indicating how certain the positive or negative classification is, based on the training data and the classification technique.

These are actually different interpretations, but not entirely inconsistent. In cases like this, the best interpretation may depend not on the algorithm you use or the assumptions behind your mathematical framework, but on your interpretation of the original training data. For instance, let's say you want to determine whether words are English words or not, and you have a labeled data set, with 1.0 meaning it's definitely

English and 0.0 meaning it definitely isn't: "the" gets a value of 1.0 while "qwxqwx" would rate something very close to 0.0.

Now, consider that one of the words is *riverrun*, which isn't in a typical English dictionary but is the first word of a famous English-language literary work. Let's say that in your data set, *riverrun* has been assigned 0.7. Leaving aside the question of whether that's an appropriate value, what does the value *mean*?

One possibility is that the person who assigned it thought, "I'm sure that this is more an English word than not, but I'm also sure that it doesn't have as much 'Englishness' as it could. It's just sort of English, or slightly English, so I'll give it a 0.7." Or maybe that person thought, "I'm uncertain as to whether this is an English word. I think that all words must be either English or not, but one can be uncertain about which category a word is in. I'm fairly sure that this one is English, but because of my uncertainty I'm going to put down 0.7."

If the entire set of training data is labeled consistently in the former way, it would be reasonable to interpret a new word that is assigned the value 0.7 as being, similarly, on the English side but only sort of or slightly English. If the entire data set was labeled by one or more people who thought like the latter annotator, the result could be interpreted as indicating some uncertainty. However, if annotators just put words into English or non-English bins, always assigning 0 or 1, it won't be nearly as easy to use the way the data was labeled to provide an interpretation of values such as 0.7.

That this classifier was trained on recent movie reviews affects not just the determination of subjectivity, but also the determination of positive and negative polarity. Such caveats are not meant to deter anyone from using sentiment analysis systems, or other types of machine learning systems, in artistic or humanistic projects. It's simply important to understand that not only the effectiveness of the algorithm but also the quality and relevance of the training data are important to getting meaningful results that lead to insights.

There are other fundamentals of natural language processing and machine learning—true fundamentals that are important for those who plan to make use of existing systems even if they do not plan on developing new ones. These include the difference between precision and recall, bootstrapping, overfitting, the difference between phrase structure parsing and dependency parsing, and ways to approach topic modeling. This chapter doesn't explain all of these, but it is meant to explain some of the essentials of both NLP and machine learning. Classification, in the specific case of sentiment analysis and more generally, is one very relevant fundamental.

[13.9]   Approaches to Text Classification

Classification (and particularly binary classification) is a relatively straightforward task, in that the system is not trying to model elaborate syntactical structures, as is done in parsing. However, the relatively simple process of classification can be very useful in text analysis and relates to concerns in the humanities. We've looked at a number of ways that a programmer, even one without extensive or professional experience, can work with text classifiers:

1. Use an already developed classifier, already trained, such as the sentiment analysis system from NLTK/TextBlob.
2. Use an already developed classifier, such as this sentiment analysis system, but retrain it using labeled training data.
3. Develop one's own classifier of a standard type (e.g., a naïve Bayes classifier) using the standard feature set for text—using word occurrences as features.
4. Develop one's own classifier of a standard type using a different, nonstandard set of features.
5. Modify classifier code so that it works for one's own classification task.
6. Write a program from scratch to do classification, implementing a standard or ad hoc classifier algorithm.

In all but the first case, training and test data is needed. In the first case, the data the system is trained on needs to be similar to the documents in the corpus that are being classified.

The last two cases are the only ones in which one actually rewrites classifier code. Programming is used in the other cases—to iterate through the corpus of documents, to train and test the classifier (in all but the first case), to develop a new classifier (in the third and fourth cases), and to specify new features (in the fourth case). But in all of these cases, the code of the classifier itself is helpfully abstracted away, bundled up and encapsulated, and the programmer uses methods to set things up. One can go look at the classifier code to see how it works, but it's not necessary for the programmer seeking to use a classifier in a program.

For most text analysis purposes, it won't be necessary to undertake 5 or 6. That work will be done by computer science and computational linguistics researchers who are developing new low-level classification techniques. There are only a few cases in which an artist or humanist might work at that level. If you want to implement a classifier on an unusual platform, for instance as part of a physical computing project, you might code a classifier from scratch. Or you might write a short and quick program, as we did, if you know that a single feature (such as our average line length) is really enough to divide a corpus of data into two classes.

In any investigation of text that involves classification, part of your challenge will be determining which of these six approaches to take. To do that, it's important to bring one's awareness of programming fundamentals to bear on the problem.

[Free Project 13-1]   Your Very Own Text Classifier
Do this project at least three (3) times. Using TextBlob, develop your own binary classifier. It's fine to use the naïve Bayes classifier. You will need to develop a labeled data set, set aside some of that data for testing, and use the rest of the data for training. And of course, in doing so, it will be necessary to determine the categories you are going to use for classification of this data.

[13.10]   Trivial Image Classification

To see that classification applies across media, and as a concluding palate cleanser, we'll turn to a very simple case of image classification—again, using one global feature to classify images that would be easily classified by a human viewer. At least, we'll use one feature at a time; we will try out a small number of features. The point here is to show that the basic concepts we saw in introducing text classification apply in image classification as well. While this will not be a sophisticated use of image classification, it will be possible to conduct this classification on an ordinary computer, without the need to use cloud services or have a high-end GPU installed. We also will prepare our own tiny data set.

We will obtain a set of ten images, each of which is a photograph of an outdoor area. Of these, five will be taken during the day and five at night. Six of these (three day and three night) will be used as training data, and four (two day and two night) will be our test, exactly as with the miniature prose/verse classification we did.

[Exercise 13-2]   Train and Test on Ten Images
Download ten outdoor photographs using the image search function of whatever search engine you usually use. For many people, of course, this will be Google, but you can also use a search engine such as DuckDuckGo, which lets you easily and directly view the image file for downloading. These images should be in JPEG format (the filenames should end with ".jpg"), half taken at night, half during the day. A watermark or small signature may not matter, but to the extent possible you should avoid using images that have text added to them, that are screenshots with nonphoto elements,

or that otherwise contain anything other than a photographic image. Aside from this, your images can be of any size and can be of your choosing.

These photographs may be taken in very different ways, with different cameras and settings, of many different sorts of places. Don't worry about whether the images you have downloaded are supposed to provide a true picture of the natural world. We are just looking to see, as with prose versus verse classification, if we can train a trivial classifier to use one feature and distinguish what a person could tell at a glance: Is the photograph taken during the day or at night?

As with your e-books, set aside three day images and three night images to use as training data. Again, you will manually set a dividing line based on the properties of this data. The other four photographs will be used for testing.

At the end of 10.4, Pixel-by-Pixel Image Analysis and Manipulation, a `redness()` function was introduced. That version of it only worked on square images 100 pixels on a side, but you were asked in the next section to generalize the function and to normalize the return value based on the area of the image. Use your completed `redness()` function as the one feature in your classifier, setting the dividing line according to the training data. Then see how well this generalizes to your test data. The process here is the same as with the ten e-books in the previous exercise, except simpler: you are using only one feature instead of two equally weighted ones.

After checking to see how redness works to distinguish daytime from nighttime photographs, develop a `brightness()` function that simply returns the average color value of a pixel—including red, green, and blue. This will be very similar to redness and simplified in some ways. Determine a dividing line based on your training data and test it on your test data.

Any image classification task will use thousands of images rather than ten. Assembling a data set, cleaning the data, and labeling it is sure to be an involved process that will likely occupy much more time than programming and training the system. The training process will not be setting one feature manually, but will involve using machine learning techniques. Nevertheless, the essential ideas of classification are visible in the manual processes you have undertaken in this chapter, using both texts and images. In addition, classifying both sorts of media shows that classification is a general technique that can work on different sorts of humanistic data.

[13.12] Essential Concepts

[Concept 13-1] Long Texts Can Be Classified
Computer programs can classify not only numbers and strings, but also more complex data—for instance, entire e-books. You should understand how to use a single rule to

classify documents and how this rule can be bundled up in a short function of your own devising.

[Concept 13-2] Classification Uses Features

Whether determined by researchers who do feature engineering or extracted automatically—as with the naïve Bayes classifier, which extracts words observed in the data to use as features—certain abstractions, features of the data, are the basis for classification. If these features do not capture critical aspects of the data, this will cause problems for attempts at classification.

[Concept 13-3] Sentiment in Texts Can Be Classified

Sentiment analysis, when used simply to determine the difference between positive and negative sentiment, provides an example of text classification. You should understand the power and some of the limitations of this type of classification and know how to use the built-in classifier.

[Concept 13-4] Classifiers Are Trained and Tested

Whether you look into the details of how a classifier works or not by reading the code—you can let that remain encapsulated, if you like—almost any classifier you use, except for a few toy examples in this book, will be trained on a substantial data set. You can do the training yourself (and should understand how to do so!) or, in some circumstances, may wish to use a pretrained classifier. When training, it is necessary to hold out some data to test the classifier; otherwise, there will be no way of estimating how well it performs.

[Concept 13-5] Classification Is a Cross-Media Technique

After understanding how text, including long documents, can be classified, you should develop an understanding of the way that images, and pairs of images and texts, can be classified as well—for now, understanding that images also have features that can be manually selected or extracted.

[14]   Image III: Visual Design and Interactivity

[14.1]

Having looked at pixel-by-pixel image processing, we'll now consider how computers can generate images in a way that relates more directly to the concepts of designers and artists—using some of the elements of visual design. Processing is built especially to help programmers sketch and explore visual designs in ways that leverage computation. In this book, we will only consider what is usually called *two-dimensional visual design*. We won't be covering 3-D graphics at all, although Processing allows programmers to sketch 3-D graphics. The extra "dimensions" into which we will extend our work are those of time (by responding to the time of day and creating simple animations) and interactivity (by allowing controls to change the visual display).

[14.2]   Drawing Lines and Shapes on a Sketch

We have already used Processing to present a very simple data visualization, in 12.3, A First Visualization in Processing. In this section, we'll return to confirm our understanding of how Processing allows us to use primitives that are motivated by graphic design. This is in contrast to directly manipulating pixels, which are the underlying elements of a computer image.

You figured out how to draw a horizontal line in Python at the end of 10.3, Generating Very Simple Images. This was a very low-level procedure, in which your program changed the color value of individual pixels. While a visual designer *can* think about an image as a grid of tiny elements, each with its own color, it is standard to think about visual design at a higher level. The elements of visual design are not just pixels, but include line, shape, intensity (how close a gray is to white or black), and color. There are other aspects of visual design often considered fundamental, but some of these, such as an image's focal point, are composite features and cannot be placed on a sketch using quantitative directions. Artists who learn to draw and paint (on paper and canvas) also explore the expressive quality of different sorts of lines. In Processing, there are ways to produce different sorts of lines, but these do not capture all the expressive sorts of ways that people draw and paint lines.

With these qualifications in mind, understand that Processing is set up to allow programmers to easily draw lines from any point to any other point on a sketch—as Python and PIL are not. Its facility for drawing lines is much better because it provides an *abstraction* of the line useful to graphic designers. You know how to have your program draw a line in Processing already—a horizontal one, at least—because you placed a line representing the mean on that sketch in 12.3, A First Visualization in Processing. Recall from there that Processing has a `line()` function. Given this, we will jump right in with an exercise, with the special requirement that you *not* consult online resources, or even the Processing reference, as you figure out this exercise.

[Exercise 14-1]   Draw a Diagonal Line

Using only this book and your previous work as a reference, create a new Processing sketch that is 800 pixels wide and 600 pixels tall. Recall that you need to place a line of code within the `setup()` function to specify this. Set the background of the sketch to be white, using a Processing function called `background()`. (Instead of looking up how this works, just try it out and see if you can figure it out.) As we did in 12.3, A First Visualization in Processing, we are seeking to simply output one static image. That means the drawing your program does can also be done with code in `setup()`, as before. Simply have your program draw a black, diagonal line from the upper-right corner to the lower-left corner.

You should use three methods here:

1. Decompose your programming work, even though this is a small and easy task, into smaller pieces. Drawing the line can be seen as the last step, for instance.

2. Look back at your previous successful project, which is very similar. Also, look back at the book to understand this project and how the necessary parts of Processing work. You don't need to use any functions other than ones you have already used, so don't confuse yourself by looking for other ways to do this right away.

3. Use trial and error as you write the program. For instance, start with a horizontal line in the middle, if you like, and modify the end points one at a time to see if you understand how your changes cause the line to change.

After you've done this, we'll move along to use Processing's ability to easily draw shapes. We'll draw circles starting with a circle at (56, 46).

From a design perspective, and taking a general view, a circle is an ellipse (also known as an oval) whose width and height are equal. Check out the Processing reference for the function `ellipse()` and write some code to quickly try it out, following the example there. Go ahead and change what you've drawn so that your circle is filled and, specifically, is all black. You do this using `fill()`, which you can also look up in the Processing reference. Finally, let's shrink our circle so that it is 20 pixels wide and (because it's a circle, it follows that . . .) also 20 pixels tall. That means that if you followed the example in the Processing reference and set the coordinates properly, your modified call to the function will be `ellipse(56, 46, 20, 20)`.

Now I'll explain how to do something a bit more interesting. We will develop a Processing sketch with a "line" of circles going diagonally from upper right to lower left. In graphic design terms, this will certainly be a line. But I'm putting "line" in quotation marks because we won't use the corresponding Processing function `line()`. We will have the program proceed from corner to corner, using iteration and drawing circles as it goes.

Go back to your 600-pixel-tall by 800-pixel-wide sketch, from the exercise you just completed. Save a new version of this code as we'll work by modifying it. Leaving everything the same, we'll get the first circle in place. By default, the first two arguments to `ellipse()` are the coordinates of the center. The center of this first circle should be 10 pixels from the top edge of the sketch and 10 pixels from the right edge. The x coordinate comes first: the right edge is 800, and 10 pixels in from that is 790. Next the y coordinate: the top edge is 0, and 10 pixels underneath that is 10. So to draw our first circle, we'll use this line of Processing code:

```
ellipse(790, 10, 20, 20);
```

You'll need to place this at the appropriate point in the `setup()` function. Also, you'll need to use `fill()`—in exactly the same way you already have—to have this circle filled in. Once you have this working, we can add the *last* circle that we wish to draw. This will be the one with a center 10 pixels from the bottom edge of the sketch and 10 pixels from the left edge. The x coordinate is first, and because we're just moving 10 to the right of 0, it will be 10. For the y coordinate, the sketch is 600 pixels high, so 590 is the point 10 pixels above that edge. The line of code is:

```
ellipse(10, 590, 20, 20);
```

Now what we need to do, pretty much literally, is connect the dots. We want to set up a for loop that starts at the coordinate (790, 10) and ends at (10, 590). So far, I haven't said anything about how the circles should be spaced. But I would definitely like them to be evenly spaced. Let's say that we'd like them to start and end in these places and to be evenly spaced. Strictly speaking, we have met that requirement already, in a *trivial* sense. There is a start circle and an end circle, with no circles between, currently. For this example, we will have our code draw twenty circles. Because the horizontal distance to be covered is 780, each step means moving the center horizontally by 780/20. The vertical distance is 580, so the vertical step is 580/20. Of course, we could compute these and stick the results into our code, but why? This would obscure how we arrived at these values, and we might make a mistake entering the computer values, and we have a computer right here to do this sort of work for us.

```
for(int circle = 0 ; circle <= 30 ; circle = circle + 1) {

  ellipse(790 - ((780 / 30) * circle), 10 + ((580 / 30) * circle), 20, 20);

}
```

Plug this in and take a look at the result. See what you need to change to draw more circles (say, forty) or fewer (say, ten).

Although I am not going to explore this in detail, if you change the number of circles to thirty, you will see that the line doesn't make it exactly to the lower-left corner. In Processing (unlike in Python 3), dividing two integers gives an integer result, trimming off a tiny bit of the horizontal change each time a circle is drawn. By the time the process reaches the lower left, the discrepancy is visible. You can explore how dividing by 0.0, as opposed to 0, yields a different result in Processing by using the println() function. Work at this further to figure out how to solve this particular problem, too.

[Exercise 14-2]   Draw Shapes of Decreasing Size

Having drawn a line of shapes, each the same size, using a for loop, you should now make two minor changes to confirm your understanding.

First, change the code so that the size of the circles is reduced. The top-right circle should be 20 pixels in diameter, as now, while the circle in the lower left should be 10 pixels in diameter, and the transition should occur smoothly.

Second, change the code so that instead of the line of circles starting in the upper right and going to the lower left, it begins on the left side in the middle of the sketch and ends on the right side in the middle of the sketch.

[Free Project 14-1]   Recreate Geometric Designs

Do this project at least three (3) times. For this project, choose a simple but interesting geometric design—for instance, one that you've found on a book cover or one featured in some organization's logo. Using shapes and lines, recreate this design. Try to do it in very few lines of code. Instead of drawing each line of a rectangle individually, for instance, you should certainly use the rect() function. But you should also do the sort of programming you did in the last exercise, using a for loop to produce several regular elements when there are such elements in your design.

[14.3]   Updating a Sketch Frame by Frame

To see how animation works, we'll begin with a very simple sort of animation. We'll draw a rectangle on the left side of the window and will then move it across the window to the right. We'll use Processing.

In creating a static image in Processing, we used the setup() function to do all the drawing work. This is a function that is run once at the beginning of a Processing program's execution. So it's just what we need for drawing a static image—but not what we need for creating a moving image.

Now that we are looking to do something different, we'll need to separate the tasks that belong in setup() from those that belong in another function, draw(). This function is called (i.e., invoked) every frame. In Processing, the frame rate can be set using the frameRate() function, which of course should usually be placed in the setup() function. If it isn't specified, the frame rate will be 60.

So, let's get going on our moving-rectangle program. To begin, we can set the display window size. Let's be square again and make it 600 × 600:

```
void setup() {

  size(600, 600);

}
```

Try this out; you'll see that a window of the appropriate size appears and that nothing else happens. To amuse ourselves, let's change the background color so it is a medium gray instead of the default light gray:

```
void setup() {

  size(600, 600);

  background(127);

}
```

Because 127 is in the middle of the possible range of intensities, 0 through 255, this is a medium value. However, whether it visually looks like the medium gray you are expecting will depend on your operating system, display settings, and the display itself.

Now, let's draw our rectangle. Add this to the existing program, below `setup()`:

```
void draw() {

  rect(0, 100, 100, 400);

}
```

If you run this, you'll see a rectangle drawn on the left side. Actually, for as long as the program is running, this rectangle is being drawn *again and again*, sixty times per second. Furiously! But you might notice that your computer isn't melting down; this redrawing is actually a pretty routine graphics task.

Let's get this rectangle moving! We need to change the *x* coordinate, the first argument. Initially, and actually always (as the program is now written), this value is 0. That puts our rectangle right up against the left side of the window. Because we want to move it along, we need to make 0 into 1, then 2, then 3; that is, we need to increment the value that now simply remains 0.

Here's how we can do it. Change the program so that it is as follows—leaving the `setup()` function alone as it hasn't changed:

```
int x = 0;

void setup() {

  size(600, 600);

  background(127);

}

void draw() {

  rect(x, 100, 100, 400);

  x = x + 1;

}
```

If you key this in and run it, you'll see that it works, after a fashion. The rectangle does move—but something weird is going on, and a large, black rectangular region is being drawn as well.

Change the line `x = x + 1;` to `x = x + 5;` take a guess about what will happen now. Then, try out the new program.

As you can see, the region behind the moving rectangle is no longer solid black. What's happening?

The `draw()` function is correctly drawing the new rectangle each time. But it isn't clearing the window; it isn't wiping out the *previous* instances of the rectangle. If we want to do that, we need to do something along the lines of redrawing the background each frame. Notice that we're going to back to + 1 instead of + 5, and then change the `draw()` function so that it is as follows:

```
void draw() {

  background(127);

  rect(x, 100, 100, 400);

  x = x + 1;

}
```

Now we have a rectangle taking flight across the screen. Let's say we want this rectangle to move three times *slower* than it does right now, even though its current speed is not tremendous. How could we do it? We might think to try adding 0.33 instead of 1 to x, but x is an integer. A better idea would be to change the frame rate. We're now running at 60 frames per second (the default), but we could set the frame rate to 20. Place the following line anywhere in `setup()`:

```
frameRate(20);
```

The glacial progress of our rectangle should show that this method works and that it slows down our animation.

[Exercise 14-3] Make a Shape Bounce

You've now made a rectangle move from the left side of the window across that window and off the right side. This is fine, but your rectangle is gone after it finishes crossing the screen. What if you wanted this rectangle to bounce off the right side and start moving left again? And then, after it did this, what if you wanted the rectangle to bounce off the left side and start moving right again? Instead of just a transient rectangle, you would have a sort of metronome. Implement this never-ending bouncing rectangle.

A hint: As you try to test this, you may find it very annoying if you start with a slow-moving rectangle and a large window. A simple way to speed things up involves increasing the frame rate.

Another hint: There's another way to speed things up. We used it already when we changed + 1 to + 5, which lengthens the steps that the rectangle takes. However, this method of increasing the speed can interact with how you test to see when the rectangle has arrived at the right edge. If you use equality to check whether x == 101, for instance, this will work fine when you're increasing x by one each time. But if you are increasing x by five each time, starting with zero, your x will skip right over 101, overshooting that value. You'll do better to use an inequality, such as x > 100, to figure out if your rectangle's horizontal position has shifted to the right of where it should be. The values 101 and 100 are examples; they aren't the particular ones you will want to use.

One final comment on changing the speed of the animation, and one that may not be evident when you work with a very simple animation such as this one: if you change the frame rate, you make a global change that speeds up or slows down everything that is being animated. That can be fine, but once you get to more elaborate animations, you may want to slow the movement of one element while leaving the rest of the animation the same. In this case, you'll have to use a different method.

[14.4]   Changing Intensity

I'll go through just one more animation trick in detail. After finishing the last exercise, you have a system in which one variable moves from zero up to a certain limit, then down again, and then up again, and so on. The variable corresponds to the horizontal position of a rectangle. But we could change a value, and move it continually up and down, to achieve a different effect.

To see how, let's focus on the *intensity* of the rectangle. This is basically the monochrome version of color. In our case, as I mentioned earlier, we have not 50 shades of gray, but 256 shades. Let's change the rectangle from 0 (black) a shade at a time up to 255 (white), and then back down, and so on. We can do this by adding a small bit of code. First, at the top, let's define a variable shade and another variable difference:

```
int shade = 0;

int difference = 1;
```

What we've done here is provide one label, or slot, that will hold the color. It's 0 right now, but it will increase a step at a time to 255 and then decrease. The other variable tells us what sort of step we're going to take. Right now, difference is set to 1, so the program will step from 0 to 1, from 1 to 2, and so on—but when we reach 255, the program will need to start going back down. Then, difference will change to -1 to enable this change in direction.

In the draw() function, after the call to background() and before the call to rect(), add this code:

```
fill(shade);

shade = shade + difference;

if (shade == 255) {

  difference = -1;

  }

if (shade == 0) {

  difference = 1;

  }
```

When the program reaches the point of maximum intensity (255), `difference` will flip to −1. When it reaches the point of minimum intensity (0), it will flip back to 1. Notice that this code will not particularly work if you change the *amount* of the step because it's checking for the shade to be *exactly* 255 or *exactly* 0.

Run the program and see how the animation looks. The effect, if not sublime, will probably be at least interesting.

If you want to start experimenting with color, just add two more arguments to `fill()`. Your arguments (as in Python) will represent the red, blue, and green color values. For instance, to have the rectangle go from black to red and back without displaying any green or blue, change the `fill()` line to:

```
fill(shade, 0, 0);
```

This method of oscillating between 0 and 255 is one way (and a fairly straightforward way) of generally moving a value from a lower bound to an upper bound. It will probably be similar to the way you made your rectangles bounce back and forth.

[Exercise 14-4]   Multiple Moving Rectangles with Color

All right, keeping going. Add several other rectangles bouncing about (get some of them moving vertically, too) and change their colors as they move. For this exercise, add three to five rectangles by duplicating code you have written and modifying it.

[Exercise 14-5]   Fifty Moving Rectangles

For this exercise, you don't need to worry about color. You can add color changes after the core of the exercise is done if you must. You can also change the rectangles to be of different sizes if you like—later. Initially, just get fifty similar-looking rectangles, each of them beginning at an independent location, to move. At the very beginning of this process, have them just move off the screen horizontally, as was done in the original animation example. Then see if you can get them to bounce back and forth.

As you can imagine from the previous exercise, it is *possible* to do this fifty-rectangle exercise by just hard-coding fifty different rectangles: copying code, pasting it in fifty times, and modifying each snippet of code. But don't do that—it helps no one! Instead, use an *array* (which in Processing is similar to a Python list). Set each of the fifty values in the array to represent the position of a different rectangle. (You can use randomness, or pseudo-randomness, to do so. Check the Processing reference pages for information on this.) Move each value so that the rectangles move. After you have them all in motion, see if you can move them according to a particular current difference or direction for that rectangle. Perhaps you need another array of fifty values for this?

If this exercise is rewarding, you can continue to get the rectangles moving at different speeds.

[14.5]   Exploring Animation Further

There are numerous examples of animation provided in Processing and made very easily accessible to you, via the File > Examples menu item. If you need to review the basics of how `setup()` and `draw()` are used to animate, for instance, there's Basics > Structure > SetupDraw. Some of these examples are illustrating more advanced facilities of Processing, but by this point in the book you should be able to locate ones that you find compelling. You can then study the code, not only by reading it and seeing what it does when run, but also by modifying it to explore how it works.

[Free Project 14-2]   Parametric Geometric Designs
Do this once for each Free Project 14-1, Recreate Geometric Designs, that you completed. For the first free project, you made your own sketch of a geometric design. In this free project, you will *parameterize* your design. That is, you'll make it dependent on values that you can change. Two parameters are required; as the "free" part of this project, you should add other parameters yourself that suit the design you developed.

To begin, add the parameters (variables) `height` and `width`, representing variation in the two dimensions of the design. These should be floating-point values, with 1 meaning no variation, 1.5 being a "stretch" of 150 percent in that direction, and .6 being a rescaling to 60 percent, compressing the design in that direction. For instance, if `height` is set to 1.2 and `width` to 1.5, the design will be enlarged, with the height being 120 percent of what it originally was and the width being 150 percent. As a result, the aspect ratio will change. This isn't supposed to be the most exciting change. It's meant to prepare you to determine at least two additional parameters.

The additional parameters you add will depend on the nature of your design. If you have a rectangular design, you may wish to allow it to skew to the left and right based on a parameter. If you produced a flower with eight evenly spaced petals, you may wish to parameterize the number of petals. Would your parameter be a floating-point value or an integer in this case? You'll need to figure such things out. Beyond the two parameters controlling how the height and width of the design will be rescaled, add at least one more thoughtfully selected parameter that suits your design.

[Free Project 14-3]   Novel Clocks

Do this project at least three (3) times. For this project, create working displays of the time using Processing. The time can be displayed any way you want, but the display should somehow change every second. You should consider ways of displaying the time that are interesting—conceptually, visually, or practically—to you. If you often talk with someone in another time zone, can you create a working clock that displays the time in both places? If you use a twenty-four-hour clock of some sort, can you display the time in a similar way? Can you have one of your clocks imitate the cycle of day and night? Does the radial display of time (as in a typical analog clock) appeal to you, or do you prefer a digital display, or perhaps an unusual watch face of the sort of you see on a binary watch? You should have some ideas in mind as you start to develop a clock of your own, but you should also be able to follow new avenues that you discover while you program. Be sure that however unusual the clocks you develop, the time can still be read from them—at least by you.

An analog-style clock is one of the examples provided with Processing: Basics > Input > Clock. If you like, you can start with this—but even if you find some nice ways to modify this, starting with this code might predispose you to producing a more conservative and traditional clock. Remember that even if you have the impulse to start from nothing and code this project, you can use this example or other code as a starting point and as something to think about. And even if you feel more inclined to begin tweaking the example, you do have the ability to start from a blank window and develop your own code. If you choose to modify the example, a good way to proceed involves stripping out various sorts of complexity and trying to understand each of these as you do so. Then you can build the example back up in your own way.

One of the nice things about this clock project is that your program must deal with data—the current time. You can't just draw a square and be done with it; you have to consider how your graphics will work in motion and at any point in the day. However, this project doesn't involve accepting input from the user, so that complexity (as interesting as it is) is factored out. There's no need right now to deal with user input to create an interesting software machine that displays the time. Interactivity is a topic, instead, for the next section.

[14.6]   Interactive Programs

A curious feature of the programs you have typed in, or written yourself, is that—if you have been following directions and have been keeping to the exercises and free projects listed—they are noninteractive. Of course, you can interact with them at a certain level by changing the code; your Python interpreter, Processing IDE, and text editor *are* interactive programs, and they allow you to change the text files that constitute other programs, and then you can run these programs again. But the programs themselves do not accept input from the user during execution and thus they are not interactive.

I discuss here some simple ways to begin accepting and processing user input. We'll get on to developing some interactive sketches in Processing, but interactive programs can be written in all sorts of programming languages, so we'll digress from visual design for a moment to demonstrate that. Let's return to Python (and our Jupyter Notebook) for a moment.

You can request input from a user, and assign what is typed into a variable, quite easily in Python:

```
name = input('Please provide your name:')

print('Hello, ' + name + '!')↵
```

Try this out. You can take this code and place it in a text file ending with ".py" if you like so that it can be run from the command line. Jupyter Notebook suffices to show you how it works, though.

[Free Project 14-4]   Conversation Starters
Do this project at least three (3) times. Develop some Python code of your own that similarly requests input. See if you can modify this "Hello NAME!" example into something a bit more interesting than this (rather insipid) greeting. You can request any typed input from the user, and rather than just printing a fixed text with a slot in it, you can generate a reply in the way you did with your random text generator (based on "Stochastic Texts") or by transforming the input. Think of this as a sort of degenerate bot or chatterbot, in the style of Eliza/Doctor and many interesting modern-day systems, that only speaks for one turn of conversation.

If you like, of course, you can extend the conversation for any number of turns—you can even develop a program that conducts a conversation of unbounded length, using an infinite loop. But for this project, start with a single reply.

[14.7] Key Presses in Processing

In standard Python, it's a bit tricky to actually register key presses, as opposed to asking for input and having someone type something followed by pressing Enter. Processing, on the other hand, is easily able to deal with keyboard input as each key is pressed. Just as there is a special setup() function and another one, draw(), Processing provides a special keyPressed() function that is called every time a key press is detected. What was typed is then available via the local variable keyCode.

What follows is the longest program in the book. Still, it is a fairly simple program and is a feature-complete generated landscape, presenting a forest with overlapping, circular canopies, seen from above. I have placed this program on my website for you at nickm.com/code/forest.pde, and you are welcome to copy it from there. It wouldn't be a bad idea to type it in, however, to continue getting a feel for programming. If you do so, remember that you don't need to type in the comments that begin with //—although it's fine to include them.

```
int dx = 0; // horizontal distance from the center

int dy = 0; // vertical distance from the center

int num = 100;

int[] treex = new int[num]; // there are 100 "trees" (circles)

int[] treey = new int[num]; // these store their x,y coords

int[] treesize = new int[num]; // diameter for each tree

int[] treealpha = new int[num]; // transparency for each

void setup() {

  size(500,500);

  // fill the position, size, and transparency arrays

  for (int i = 0; i < num; i = i + 1) {

    treex[i] = (int) random(width * 2);

    treey[i] = (int) random(height * 2);

    treesize[i] = (int) random((width + height) / 8);
```

```
      treealpha[i] = (int) random(128);

  }

}

void draw() {

  background(255); // white background

  // move the canvas by the current dx, dy amount

  translate(dx - (int) (width / 2), dy - (int) (height / 2));

  for (int i = 0; i < 100; i = i + 1) { // go through all 100 elements

    fill(0, 0, 0, treealpha[i]); // black, transparent interior

    stroke(0, 0, 0, treealpha[i]); // so is the outline

    ellipse(treex[i],treey[i],treesize[i],treesize[i]);

  }

}

void keyPressed() { // have arrow keys change dx, dy

  if (key == CODED) {

    if (keyCode == DOWN) {

      dy = dy - (int) (height / 10);

    }

    if (keyCode == UP) {

      dy = dy + (int) (height / 10);

    }

    if (keyCode == RIGHT) {

      dx = dx - (int) (width / 10);

    }

    if (keyCode == LEFT) {

      dx = dx + (int) (width / 10);

    }

  }

}
```

This program generates a different abstract forest each time it is run. The view is initially in the center of the forest (seen from above), but by using the arrow keys, whoever is running the program can navigate around. Randomness (or the computer's approximation thereof) is used at the beginning, to determine the location, size, and transparency of each tree. After that there is nothing random about how the program works. When the arrow keys are pressed to change the view and then change it back to what it was before, the exact same thing will be shown. This landscape is *generated* in the same way the text from our random text generator was. We didn't manually place the trees at fixed locations but wrote a program to do that, in this case using Processing's random() function. It is *navigable* because of how the arrow keys allow us to see more than the initial view.

[Free Project 14-5]   Create a Navigable Generated Landscape
Do this project at least (2) times. It's fine if one is a modification of my code. Write something from scratch as well. Use the previous code as a starting point and develop some modified code from it. Initially, it may help to proceed incrementally by changing the code, better understanding it, and adding one element at a time. For instance, it's possible to change the controls. As a first step, you could allow WASD-style navigation instead of (or in addition to) navigation with the arrow keys. Implementing this will allow you to see how keyCode works. You can also use a mouse or trackpad as an interface. See mouseX and mouseY in the Processing reference, to begin with. You detect interface events such as dragging with the mouse, too. A good starting point for this would be mouseDragged() in the Processing reference.

After you are familiar with how this code works, develop your own unique and systematic virtual, navigable space. This means simulating a space that is larger than the Processing window. You can choose how that is done; your program doesn't have to use the top-down view provided in this example. It also means developing a control scheme, which doesn't have to be based on the arrow keys (or, for that matter, mouse dragging). Finally, it means generating the landscape using code. Your landscape doesn't have to be random: a highly regular cityscape or cemetery or quincunx would be fine as well and can be accomplished with iteration. But if you have a list of points and are just asking the program to connect the dots, you aren't generating a landscape; your program is just serving as a plotter for prespecified data.

Because you have this sample code, you can use it as a scaffold, even if you plan on replacing all of it in your finished project. Some major components here are the interface, the generation of a landscape, and having only a portion of that landscape shown. Don't try to completely rewrite all three at once. Instead, change one of the aspects and use the provided code to help you see that your change is working well.

This free project puts together the new topic of interactivity with some generation ideas we dealt with earlier from the standpoint of language. It also offers a chance to deal with windowing (in the most general sense) and keeping track of things that are out of view, for now, but which may be displayed later.

Processing allows you to easily create a sketch that you can show in class or present in exhibit. However, it is no longer easy to share Processing work on the Web because of security issues with the whole concept of Java applets and running Java within the browser. To easily share Processing work on the Web, you can write code in a system called P5.js. The P5.js project, which has its home page at p5js.org, was founded is and led by Lauren McCarthy. It is a JavaScript framework that provides Processing-style functions within JavaScript programs. There are a few additional complexities to using P5.js, so I focused our work in this book on good old Processing and today's version 3 of it. But many of those who learn from this book will probably want to combine their knowledge of JavaScript and HTML pages (e.g., from 3.3, Quick and Easy Modifications) with what they know about Processing and learn to work in P5.js.

[14.8]   Essential Concepts

[Concept 14-1]   Abstractions Facilitate Visual Design
Programming languages provide ways to process images that in some ways relate to the computer's underlying image representation (amounts of red, green, and blue that are between 0 and 255) and in some ways relate to higher-level design thinking (being able to specify red as a color and draw lines and shapes, rather than turning on and off individual pixels). Understand that there is usually some balance between these and that useful abstractions will help designers and artists accomplish visual work.

[Concept 14-2]   Drawing in Time Produces Animation
The contents of draw() allow for animation to happen as different things are drawn on different frames. The function has to be written, somehow, so that what is drawn changes from frame to frame. This can happen by saving state in a variable and incrementing or otherwise changing it, or it can happen by introducing a dependence on time or user input.

[Concept 14-3]   Interactivity Is Accepting and Responding to Input

Understand the basic ways to take input from the keyboard in Python and Processing and how input can either be accepted in a "turn-based" way, as when a name or other string is provided, or a "real-time" way, as when input controls a dynamic animation or allows one to navigate around a space.

[Concept 14-4]   A Window Looks onto a Virtual Space

Input can be used to move a window around and view different parts of a virtual space. To do this requires a model of the larger space that is not directly seen. Understand the principle here and its applicability to areas of a Web page or word processor document that are viewed by using scroll bars to move vertically and horizontally.

[15] Text III: Advanced Text Processing

[15.1]

This chapter covers some more sophisticated types of text processing, both the humanistic (analytical) and artistic (generative) types.

The first consideration is, once again, the seemingly simple but actually rather tricky problem of counting words. This time, word counting is done using TextBlob. The discussion then covers how to gather word-based statistics (such as average word length) and how to count words of certain types, accomplished with a very reliable type of complex natural language processing software, a part-of-speech tagger. (As with all such systems, it works best in processing the specific types of data used to develop it.) Segmenting a document into sentences is then covered.

After these text-analysis methods, the discussion continues with some ways to use simple word lists and an excellent handcrafted lexical resource, WordNet. The last part of the chapter deals with language generation, which was a very early topic, first encountered in chapter 3, "Modifying a Program." The advanced discussion of language generation covers how to develop "cut-up" texts, how to define grammars for generation, and how language generation can be recursive.

These are only a few directions in which to begin to explore text processing, but they are representative of different capabilities and approaches. This book doesn't dip into how to relate lexical (written) language to phonetic (spoken and heard) language, although there are plenty of ways you can do that as a programmer. To go beyond our topics here in terms of analysis, chunking text, parsing it using dependency or tree parsers, and recognizing named entities are a few of the other interesting topics in natural language understanding. Readers should be well-prepared to explore those after this chapter. There are many other possibilities for creative text generation as well, of course.

[15.2]   Words and Sentences

When we speak of *words* to discuss something such as the number of words in a text, there are at least two things we might mean. Consider this well-known text, the first sentence of *A Tale of Two Cities* by Charles Dickens, available in approximately the same form from Project Gutenberg:

> It was the best of times, it was the worst of times, it was the age of wisdom, it was the age of foolishness, it was the epoch of belief, it was the epoch of incredulity, it was the season of Light, it was the season of Darkness, it was the spring of hope, it was the winter of despair, we had everything before us, we had nothing before us, we were all going direct to Heaven, we were all going direct the other way—in short, the period was so far like the present period, that some of its noisiest authorities insisted on its being received, for good or for evil, in the superlative degree of comparison only.

In the standard sense of a word count, it would be correct to say that this sentence has 119 words. This is a view of words as *tokens*. There are 119 individual sequences of letters, separated by whitespace and punctuation, laid out one after the other. But it would also be correct to say that the sentence has fifty-eight words, if we are speaking of unique words. This is the view of words as *types*. There are eleven occurrences of *was* in this sentence, and so those occurrences account for one type and eleven tokens. When people speak of a *word count*, they are usually taking a view of words as tokens. In particular situations in which it may be unclear, as when one begins to discuss word usage, frequency, and so on, explicitly using the terms *types* and *tokens* can be very helpful.

In natural language processing, an important early step is dividing a text into words (in the sense of tokens). It's not just important for purposes of counting words and making comparisons. If one is interested in assigning each word a part of speech or in parsing sentences to determine their grammatical structure or in classifying text according to the words used or in most any other type of language processing, it's necessary to first determine what the words are in the text. So some thought has been given to this preliminary type of processing, this type of segmentation of a text. This process is called, for reasons that should not surprise you, *tokenization*. The text is not being divided into types, after all.

We have written code to segment texts into words already, but we can use a well-developed method of tokenizing text by calling the built-in tokenizer that TextBlob provides.

Fire up Jupyter Notebook. To get started, we'll need to `import` the class we're going to use.

```
from textblob import TextBlob↵
```

Then, we can create our first TextBlob, this time by just providing a short sentence as an argument to the `TextBlob()` constructor:

```
gravity = TextBlob('A screaming comes across the sky.')↵
```

The variable `gravity` does not hold a string, even though we used a string to create the value it does hold. This variable holds a TextBlob object, and we can call methods and view attributes of such an object.

```
gravity.words↵
```

This is asking for the `words` attribute of `gravity`. Notice that the result is exactly the same as if we had typed in:

```
TextBlob('A screaming comes across the sky.').words↵
```

This should no longer be a surprise, but still, it's good to reassure yourself about how things work now and then. The result that you get is a special kind of object, too—*not* a list, but a WordList object. As with a TextBlob, you can do some additional things with this object because it has methods and attributes that are not provided for lists. However, you can also treat it like a list. Try, for instance:

```
len(gravity.words)↵
```

As well as:

```
gravity.words[:3]↵
```

And, try this one also:

```
sorted(gravity.words)↵
```

Now, you might complain that this attribute is, confusingly, called *words* when it could have been called *tokens*, to let everyone know that it is word occurrences we are finding. Try this:

```
gravity.tokens↵
```

You can see from this that the sentence has one more token than it does words (or words viewed as tokens). The final piece of punctuation is included in the list of tokens. So as an additional wrinkle, not only can words be thought of in terms of types or tokens, but when a text is divided into tokens, these can include some tokens that are not words.

You can see from the following example that words and punctuation are not separated according to any of the lexical rules we discussed earlier:

```
TextBlob("Can't touch this.").tokens↵
```

Notice that this string begins with " and ends with ". I used double quotes specifically so that it would be easy to include the apostrophe in Can't. Also, notice that the result is, well, a bit unusual from a purely lexical standpoint. Does this have something to do with the piece of punctuation in "Can't"? It can't, actually, as this example will show you:

```
TextBlob('Cannot touch this.').tokens↵
```

The tokenizer that is being used here models words in yet another way. It's a way that is well-grounded, linguistically. But we shouldn't expect it to match the word count that a text editor or word processor does.

[15.3] Adjective Counting with Part-of-Speech Tagging

Let's create a TextBlob with a slightly longer sentence for this next task, by typing all of this on one line:

```
pride = TextBlob('It is a truth universally acknowledged, that a single man in
possession of a good fortune, must be in want of a wife.')↵
```

We're looking for a way to identify the adjectives in this sentence (and any sentence) so that we can count them. At this point, though, we do not need to determine the grammatical *structure* of the sentence. We're not going to draw one of those sentence diagrams or determine which noun is the subject and which the object. We just want to be able to tell which words are which part of speech. A system that can do this task is a *part-of-speech tagger,* also called a POS tagger. Such systems are highly accurate, if the text being tagged is similar to the text used to train the system. And there is a default tagger (just as there is a default tokenizer) built into TextBlob.

```
pride.tags↵
```

As before, if you need to install an NLTK resource, just do exactly what it says there in Jupyter Notebook and try this again. You are to type this in yourself and see what results, but for purposes of discussion, here is what the beginning of the output should look like:

```
[('It', 'PRP'),
 ('is', 'VBZ'),
 . . . .
```

Now, to figure out what exactly this data is, we can start by determining its *type*. See if you can clearly state the type of this value. Hopefully, you can; in any case, continue reading.

This is not a number (an integer or a floating-point number), and, although there seem to be some strings in there, the whole thing is not a string. It begins with [and ends with] and has several things delimited by commas. So at the highest level, we are dealing with a type of data we are quite familiar with: a list.

So what is this a list of? It's a list of things that begin with (and end with) and have a comma in the middle. Those things are tuples, and specifically, pairs. And what are they pairs of?

Things that begin with ' and end with ' and contain sequences of characters. These are strings.

So, the type of `pride.tags` is: a list of pairs of strings.

You can see from that list that the first element, the first string, is in each case a word from the text we provided, the first sentence of *Pride and Prejudice*. The second element, the second string, is the part-of-speech tag.

These tags are based on the ones used to annotate the Brown Corpus, or, in full, the Brown University Standard Corpus of Present-Day American English. This influential linguistic resource was compiled in the 1960s and had about one million words in many different genres. To label each token with a part of speech, eighty-seven tags were developed.

The Penn Treebank project, which ran from 1989 to 1992, resulted in a corpus of more than 4.5 million words that was annotated for syntactical structure. The tags used for this project were from a reduced set; there were forty-eight, of which twelve were used to label symbols that were not words. This tag set is the one used by both of the taggers included in TextBlob.

You can pick out, just from this sentence, what several part-of-speech tags are. *NN* means noun, *IN* means preposition, and it looks like *JJ* means adjective. So let's count the adjectives. How should we proceed, using the sorts of techniques that we've already discussed?

The most straightforward way is to set a counter variable to 0 and iterate through the list, adding one to the count each time an adjective is encountered:

```
count = 0

for (word, tag) in pride.tags:

    if tag == 'JJ':

        count = count + 1↵
```

[Figure 15-1]
Part-of-speech tagging associates tokens with grammatical categories. Notice that *'s* is assigned its own tag here, as it is a contracted form of *is*. Here, the categories indicated by tags are *DT* (determiner), *VBZ* (third-person "be," singular, present), *JJ* (adjective), and *NN* (noun, singular or mass).

Wrap that up in a function called `adjs()`, using your knowledge of how to write functions. Where `pride` occurs, you'll do well to replace that variable name with a generic name for a TextBlob and then make that the argument to the function. You'll want to return the value of `count`.

Check out your function `adjs()` by taking a look at how many adjectives are in the first sentence of *Pride and Prejudice*:

`adjs(pride)` ↵

And, for good measure, check out how many your code counts in the first sentence of *Gravity's Rainbow*:

`adjs(gravity)` ↵

Keep this session open with this function in memory, as we'll make further use of `adjs()` before too long.

[15.4] Sentence Counting with a Tokenizer

The tokenizer not only segments text into words and other symbols, but also into sentences. If you try doing this with a very naïve method, such as splitting the text every time you encounter a period, it may work on certain texts. But for others that contain *Mr.* or *Ms.* or *Dr.*—or any other sorts of abbreviations ending in periods—it won't work. And if you have an ellipsis (. . .), the simplest sort of rule will split the text into three sentences at that point.

Let's load up the text from the Universal Declaration of Human Rights again, this time in a TextBlob, with both sentences typed on a single line:

```
hr = TextBlob('All human beings are born free and equal in dignity and rights.
They are endowed with reason and conscience and should act towards one another in
a spirit of brotherhood.')↵
```

Now, as a follow-up to checking the words attribute, try checking the sentences attribute, as follows:

```
hr.sentences↵
```

Again we have a list, and this time it's a list of Sentence objects. As we've seen in other cases, we can treat Sentence objects as if they are lists, for instance:

```
len(hr.sentences)↵
```

[15.5] Comparing the Number of Adjectives

Now let's load up two novels, one after the other, and see how many adjectives are in each. We can normalize our result by determining the percentage of adjectives: the number of adjectives we find divided by the number of words overall.

Let's use *Pride and Prejudice* (1342-0.txt) and *Moby-Dick* (2701-0.txt), the plain-text versions that are both easily available from Project Gutenberg. We'll use the same method as before to read in a file. As before, the files must be located in the same directory where Jupyter Notebook is running. Because you have already prepared text files, removing front matter and end matter from them, in exercise 13-1, Train and Test on Ten Books, you should use those cleaned versions of the Project Gutenberg files. In the example that follows, I use pride.txt to indicate such a file. In the first line that follows, remember to add the argument encoding='utf-8' if you are using Windows:

```
source = open('pride.txt')

pride = source.read()

source.close()↵
```

Now, do the same thing for *Moby-Dick* (also using a cleaned version of the file), placing the text in variable moby. With that done, you have the full text of both novels assigned to the variables pride and moby.

Check out the very beginning—the first hundred characters, for instance—of each of these long strings. You can do this, of course, by taking a slice of these strings.

Next, create TextBlob objects:

```
prideblob = TextBlob(pride)↵
```

And the same thing for mobyblob, which will represent the TextBlob of *Moby-Dick*. At this point, we're ready to see how many adjectives there are. We just should invoke adjs() with prideblob, and then mobyblob, as an argument. Go ahead.

Having done this, however, we should soon see that we really want to normalize over the total number of words, to determine the ratio of adjectives to words overall. If a book half as long had as half as many adjectives, that means the text is not different but similar to whatever it is being compared to.

So we will want to divide the number of adjectives in, say, *Moby-Dick* by the total number of words, which is the length of mobyblob.words. This, of course, is determined using len():

```
adjs(mobyblob) / len(mobyblob.words)↵
```

If you check this ratio for both texts, you'll see that one text has a percentage of adjectives that is almost one and a half that of the other. Does this mean much to you? Does it indicate a difference in narrative style? At the very least, it shows that the ratio of adjectives *does* indeed vary. If every novel had about the same percentage of adjectives, it almost certainly wouldn't be interesting to track and compare them. But future work along these lines, embracing a larger corpus of novels, has the potential to show interesting patterns, whether or not there is anything to be gained by comparing these two particular novels in this way.

[15.6] Word Lists and Beyond

An inflected word list contains not just the base forms of words, which you would find listed in a dictionary, but also all the other surface forms of the word. In English, such a list will have plurals (*children* and *houses* as well as *child* and *house*), different conjugations of verbs (all the forms of *be* and all other regular and irregular verb forms of other verbs), and comparatives (*faster* and *fastest* as well as *fast*). *The Official Scrabble Player's Dictionary* includes such a list, as all forms of English words are legitimate in Scrabble play, although this dictionary only includes words of two to eight letters. Such lists are typically used by spelling checkers because they cover all forms of words.

To return to our type/token distinction, the inflected word list can be seen as a list of *types* (rather than tokens) that are not proper names and are found in English documents. Each one occurs only once, no matter how frequent or infrequent it is in the corpus considered.

There is no neutral way to gather such a list. Whether it is harvested from Web pages or assembled by hand with editorial care, decisions must be made. In the case in which the list is built automatically, the decisions include which pages to use, which to count as English (or whatever language is the focus), and whether documents of other formats (such as PDF) will be used in addition to HTML pages. For instance, *seo* may be identified as a common word in a corpus of Web documents, given that it is an initialism that stands for *search engine optimization*. It would be difficult to argue that it is somehow objectively in the top ten thousand words of English as it is used today in speech, and it is unlikely to be a common word in a corpus of old public-domain books.

It is of course possible to assemble a set of Web pages such that an unusual term like this, or essentially any word that occurs on the Web, is in the set of ten thousand most frequent words. Take a single page that is less than ten thousand words long, for instance, as your *complete* set of data: every word on that page is now in the top ten thousand, according to the process you used. The top words always depend on the corpus of text being used. If the pages one uses are out-of-copyright books from Project Gutenberg, very recently coined terms will be unrepresented. The pages of a giant online product catalog will have a different lexicon from a forum in which people discuss rare diseases, and that forum will have a different lexicon from one in which people discuss the ideal way to submissively serve drinks to one another.

While there is no inflected word list that is simply correct, such lists can still be a help. I sometimes use an inflected word list in writing constrained poems, and even though it lacks *blog* and *mosh*, it is very useful. In terms of the analysis of language, to choose a very simple example, if you start with a word list that reflects a reasonable idea of the standard English lexicon, you can see what percentage of words that people use on different pages are not in this lexicon.

Here's a quick project that illustrates the use of an inflected word list. We'll seek reduplications in such a list. As introduced in exercise 8-7, Reduplications, a reduplication of the strictest sort is a word that consists of two identical strings, one after the other. For instance, *tutu*. Reduplications as we're discussing them don't have to be a single syllable followed by the same syllable: *hotshots* counts, as it is a repetition of the string *hots*. Notice that if we were using an *uninflected* word list, without plural forms, we wouldn't detect *hotshots*.

To start, you should locate or generate an English-inflected word list that is contained in a text file, with one word per line. Search for and download such a list. Save it to where you are working (the explore folder) and call it English_word_list.txt. If searching for an inflected word list isn't effective, try searching for a spell-check dictionary. Or you can generate a suitable word list using a tool such as the one at app.aspell.net/create. Open up the file and read it in:

```
source = open('English_word_list.txt')

words = source.read().split()

source.close()↵
```

There's a little trick in the second line. `source.read()` reads everything in the file into one giant string. But another method is applied after that: `split()`, which breaks this giant string apart into individual words, splitting it on the whitespace (newlines, in this case) and discarding all of that whitespace. That second line does the same thing as `words = source.read()` followed by `words = words.split()`, but more concisely. Now, check out the first few entries in the list:

```
words[:5]↵
```

Hopefully those look like the first five words in the file. Next and finally, we want to go through the entire word list (iterate through it, as we've now done many times) and check to see if the word we're looking at is a reduplication. It's a fairly simple process, which can be tested in one line. Here is the skeleton of the three lines of code that are needed:

```
for _____ in _____: # Iteration through the word list

    if _____: # Conditional testing to see if we have a reduplication

        print(____) # Only if we do, print the word↵
```

I've used `print()` in this example because we are testing out a short bit of code that hasn't yet even been made into a function—so there's no way to return a value. Now that you're comfortable with the difference between returning a value and displaying it, just try to develop some working `print()`-based code; later, it can be improved and made into a proper function.

Filling out the first line to do iteration shouldn't be a problem. Filling out the last line should be simple, too. So do those, then start to work on the second line. How can you determine if a word has a first half that is the same as its second half? You should refer to your answer to exercise 8-7, Reduplications, of course, but after you have developed a way to do this, consider my answer.

The expression I used is `w[:len(w)//2] * 2 == w`. This is one of several expressions that will work. First off, I called the variable that was holding the individual words `w`. You may have called it something else, such as `word`; it's up to the programmer. Then, I took a slice of `w` from the beginning up to the halfway point, which as a floating-point number is `len(w)/w`. However, we need an integer to be used in slicing this string, so we use integer division: `len(w)//w`. Then, I multiplied this string by 2. We've of course seen that string multiplication is implemented in Python and that this is a legitimate way to double a string, as in the function `double()`. So that gives us the left-hand side of the equation: the first half of a string, doubled. For `'cowboy'`, this will give us `'cowcow'`, for instance. What remains is to check to see if this is the same as the word itself. For `'cowboy'` it isn't, but for `'tutu'` it will be.

There are plenty of other ways to check for reduplications. Addition could have been used instead of multiplication. A string's first half is *always* equal to its first half, so really we could just check to see if it has a right half equal to its left half. And so on.

One thing that may seem odd is that in the program structure I outlined and in my solution, I didn't check to see whether the word has an odd or even number of characters. A word with an odd length can't be a reduplication as we have defined it; shouldn't we check to see if we have an odd-length word and only proceed if the word is of even length? That's a legitimate approach, but if the conditional is constructed properly, it's not necessary. When we take the first half of a seven-character word, using `w[:len(w)//2]` or something like it, we get only three characters. When we double those three characters and get a six-character string, there's no way it will ever equal the original seven-character string, so we won't ever be mistaken.

You could *possibly* wind up with a comparison that doesn't work, though. If you decided to somehow compare the first `len(w)//2` characters of a string with the last `len(w)//2` characters—this isn't the easiest thing to do in Python, but it's possible, of course—and you didn't check before or after to see that the string was of even length, you might identify *ingoing* as a reduplication when it really isn't one. That word has an *o* in the middle and everything before it equal to everything after it. In this situation, as in many situations, finding a simpler solution is best.

To understand more about standard vocabularies and their usage, you can use other lists as well beyond the inflected word list that you found. If you want to recognize the use of standard place names, for instance, you can use a gazetteer, a resource that lists such names. Of course, looking for such names by simply matching strings is not always straightforward: *Providence* could be the capital of Rhode Island or a divine quality, to say nothing of whether or not *Intercourse* refers to the town in Pennsylvania. But so long as the limitations of such searching are kept in mind, researchers and artists can make good use of such a list and other similar lists.

[15.7]   WordNet

WordNet is a pretty amazing resource—a free hyperdictionary and hyperthesaurus that has been mainly used by researchers, but is in some ways has succeeded in the dictionary realm as Wikipedia has in the realm of the encyclopedia. The analogy shouldn't be overstated: WordNet is not a massively participatory project and is no replacement for the historical and etymological information in, say, *The Oxford English Dictionary*. It was created at Princeton University as a hierarchical (in many ways!) and carefully curated project. WordNet is, however, a rich and powerful lexical database, it is free as in free software, and at this point (having installed TextBlob) you have it installed on your computer.

Typically we imagine that the core information in dictionaries consists of definitions and that synonyms are the core information in a thesaurus. Actually, the thesaurus is a much more radical book; what people often think of these days is the synonym finder that is a thesaurus "in dictionary form." *Roget's Thesaurus*, published in 1852, actually consisted of a hierarchical categorization of concepts based on Leibniz's philosophy. Books of this sort are still published (*Roget's International Thesaurus*, for instance), and WordNet is among other things more a thesaurus in this sense than a synonym finder.

There is good comprehensive documentation of WordNet online, and, speaking of Wikipedia, the Wikipedia page is not a bad place to learn the different types of relationships that are modeled in WordNet. I'll describe a few of these and why they are interesting or complicated for artistic and humanistic programmers.

You can "manually" look up information about certain words in WordNet using the online interface: wordnetweb.princeton.edu/perl/webwn.

This is a fine way to start learning the structure of WordNet. However, you don't have to be online to use WordNet; you can find the same information using the Python interpreter, either by typing a line of code or by writing a program to use the database. To investigate the synonyms that the word *bank* has, for instance:

```
from textblob import Word

bank_word = Word('bank')

bank_word.synsets↵
```

What you see there isn't a list of synonyms, but a list (a long one) of *synsets*, sets of synonyms that each pertain to a different sense.

If you start with most common words, you'll need to distinguish the *sense* of that word that is intended. There are ten senses of the noun *bank*. The first two are enough to provide a classic example of polysemy. The first (according to the definitions that you can see using `bank_words.definitions`) is "sloping land (especially the slope beside a body of water)," and the second is "a financial institution that accepts deposits and channels the money into lending activities." When you select a particular sense, you are also choosing a synset, a set of particular words or phrases (called lemma names) that for most all practical purposes mean the same thing and relate to the specific sense. The first sense of *bank* is in the synset containing simply the lemma name *bank*, while the second is in the synset containing *depository financial institution, bank, banking concern*, and *banking company*.

```
bank1 = bank_word.synsets[0]

bank1.lemma_names()↵

bank_word.synsets[1].lemma_names()↵
```

If I had wanted to do all of this consistently, I would have assigned `bank_word.synsets[1]` to the variable `bank2` and then checked to determine the value returned by `bank2.lemma_names()`. The same thing can be done either way, and here I've done it both of these ways to show that both are possible.

While these two sets of words and phrases happen to be different, in general the lemma names are not enough, by themselves, to disambiguate what sense of a word is being indicated. Sense 3 of *bank*, "a long ridge or pile," has a synset with just the lemma name *bank* in it, just like the synset for the first sense. So if you want to know if two synsets are the same, you should check to see if the two Synset objects are equal, not if they have the same lemma names as returned by their `lemma_name()` methods.

```
bank3 = bank_word.synsets[2]

bank3.lemma_names()

bank1.lemma_names() == bank3.lemma_names()↵

bank1 == bank3↵
```

By the way, earlier we got a long list of synsets that contained both noun synsets (such as `'bank.n.01'`) and verb synsets (`'bank.v.01'` and so on). That makes sense, as there is a noun *bank* and a verb *bank*. But we probably want to restrict our results to the appropriate part of speech. To do that properly, we should import a special constant that indicates a noun. Then, that constant can be used with the `get_synsets()` method to obtain just the synsets we're interested in:

```
from textblob.wordnet import NOUN
```

```
bank_word.get_synsets(NOUN) ↵
```

You can probably guess how to get just the verb synsets of a particular Word object. Give it a try and see if you're right.

Having understood the basics of synsets, let's look at one type of WordNet relationship that isn't explicitly found in most dictionaries or other reference books. This is a relationship of generality and specificity. *Sedan* is one particular more specific term for a car, while *motor vehicle* is a more general term. Let's find the appropriate sense of *car* and then use WordNet to determine all the terms at the next level of specificity and the one term at the next level of generality:

```
car_word = Word('car')
```

```
car_word.get_synsets(NOUN) ↵
```

Synsets are ordered by how common each sense is, so we can expect that the automotive sense (as opposed to, say, the use of the term to indicate a chariot) will be early in the list. It is, indeed, first:

```
car1 = car_word.get_synsets(NOUN)[0]
```

```
car1.definition() ↵
```

So, with the right synset easily labeled as `car1`, we can check to see what is one level of generality up from *car* in the automotive sense:

```
car1.hypernyms() ↵
```

```
car1.hypernyms()[0] ↵
```

```
car1.hypernyms()[0].lemma_names() ↵
```

The lemma names for this word are *motor vehicle* and *automotive vehicle*, which both sound like reasonable terms to encompass *car* and similar entities. To see what is one level of specificity down from *car*, try the following:

```
car1.hyponyms() ↵
```

You can see that there's a synset named *sedan* in there along with many others. One of the interesting things that WordNet can assist with is changing a text to be more general or more specific. You can also see how close or far away two different nouns are, using the `path_similarity()` method:

```
river = Word('river').get_synsets(NOUN)[0]↵
```

Here I have just grabbed the first sense of the noun *river*. I intend what I think is the most com-mon sense, but I had better check to make sure it is really the sense I intend:

```
river.definition()↵
```

```
enigma = Word('enigma').get_synsets(NOUN)[0]
```

```
enigma.definition()↵
```

```
river.path_similarity(enigma)↵
```

```
river.path_similarity(bank1)↵
```

So, according to WordNet, a river is more similar to a bank (in the first sense of that word) than it is to an enigma. That, at least, seems right. To be sure you understand how this works, you should also check how similar an enigma is to a bank (sense 1) and how similar a car (sense 1) is to these three senses of words.

Because of how synsets are listed, you can often grab the first sense of a word and be right about your guess. However, WordNet's `path_similarity()` method doesn't compute the distance between two tokens in the text you are analyzing. It determines how similar two *senses* of words are. Consider what would happen if you wrote a program that happened to choose the token *car* when it refers, in the text, to a cable car or a train's dining car. If your program proceeded to grab the first sense of *car* and selected a hypernym to find a more general term, and it then proceeded to use the first lemma name, you would end up with *motor vehicle*, which is not what a cable car or dining car are.

WordNet is an excellent resource, and this quick introduction just points the way to the many facilities it offers. It's important to remember that while it offers a richly interconnected database of the different senses of English words, it doesn't determine for you what the words in your text mean.

[Free Project 15-1]   Creative Conflation

Do this project at least two (2) times. Develop a program in the vein of *Alice's Adventures in the Whale*, discussed in 9.4, Finding the Percentage of Quoted Text. Your program should somehow conflate two novels or other prose books. Obtaining these from

Project Gutenberg will probably be easiest and will mean that the format of the two books will be similar. Although you can do a conflation using only the techniques discussed in chapter 9, "Text II: Regular Expressions," here you are to join your understanding of regular expressions to the capabilities provided by the TextBlob library. You can replace text from one book with that of another, if you like, or you can interleave text from the two books. Using TextBlob, you might make fine-grained replacements—for instance, taking the adjectives you find in one book for use in another. There are other many methods as well: be creative as you conflate!

[15.8] Automated Cut-Ups

In "In the Cut-Up Method of Brion Gysin," William S. Burroughs described how to create a new text by literally cutting a sheet of paper and rearranging the pieces (Burroughs 2003). The idea was not original to him or Gysin; Burroughs points back to Tristan Tzara, a surrealist, but there are many earlier examples. One poetic tradition of cutting up and rearranging text is that of composing a cento (the word means *patchwork*), which has been around for centuries.

With some programming background, you can learn to cut up a text in new, surprising, and compelling ways. What I am going to discuss here is how to take a text that is already "choppy," or segmented into different phrases, and shuffle these around to explore how the ordering of the phrases matters in the original work—and to make a new work, which may or may not be interesting on its own. You don't have to use the cut-up method in this way. When cutting into paper, you can cut through individual words and letters if you like and make a collage that is difficult to read. But to begin I'll explain how to create this more conservative type of cut-up automatically, using a little bit of Python programming.

As a first example, we'll start with the second part of Gerald Manley Hopkins's poem "Pied Beauty":

> All things counter, original, spare, strange;
>> Whatever is fickle, freckled (who knows how?)
>>> With swift, slow; sweet, sour; adazzle, dim;
> He fathers-forth whose beauty is past change:
>>> Praise him.

Let's represent the different things Hopkins mentions in a Python list—specifically, a list of strings:

```
things = ['counter', 'original', 'spare', 'strange', 'fickle', 'freckled',
'swift', 'slow', 'sweet', 'sour', 'adazzle', 'dim']
```

We can represent the five lines of verse, with slots ready to receive these words, by writing a function that accepts a list of words and returns the verse. Now that we're nearing the end of the book, I am going to assume you have a feel for typing in code and don't need to type in a large awkward block of code for the sake of practice. So, *don't* type in the following first attempt at a function to piece those words back together into Hopkins's verse. I will leave off the ↵; I did the typing for you, so you don't need to suffer.

```
def pied(words):

    verse = (
        'All things ' + words[0]  + ', ' + words[1]  + ', ' + words[2] +
        ', ' + words[3]  + '; ' + '\n' +
        ' Whatever is ' + words[4]  + ', ' + words[5] +
        ' (who knows how?)\n' +
        '    With ' + words[6]  + ', ' + words[7]  + '; ' +
        words[8]  + ', ' + words[9]  + '; ' + words[10]  + ', ' +
        words[11]  + ';\n' +
        'He fathers-forth whose beauty is past change:\n' +
        '
        '
```

```
print(pied(things))
```

This actually works, if exactly this text gets typed in, but it's pretty much a brute-force approach to putting twelve words back into some lines of verse. Each word in the list gets placed back according to its number. There are many, many opportunities to make mistakes in this messy block of code. So here's a different approach. This time, you *should* type in the code that follows, as usual, although you are allowed (and encouraged) to copy the lines of the poem from an online source and replace the "things" with underscore (_) characters. That is, go and find the text starting with "All things . . ." and ending with "Praise him." and go ahead and place it between the beginning """ and the ending """. Then, edit that text to change words to slots.

```
def beauty(words):↵

    from re import sub↵

    verse = """All things _, _, _, _;↵
Whatever is _, _ (who knows how?)↵
    With _, _; _, _; _, _;↵
```

```
He fathers-forth whose beauty is past change:↵

    Praise him."""↵.

  for word in words:↵

    verse = sub('_', word, verse)↵

  return verse↵
print(beauty(things))↵
```

This is much easier to put together. It's a bit shorter, for one thing: 316 characters as opposed to 518. But it's also more principled and there are fewer opportunities to make line-by-line errors. This code uses iteration to go through the list words one at a time, and each time it performs a simple regular expression substitution. However, once you do actually type it in and run it, you'll notice that this version doesn't work. Try it and see what sort of failure happens.

The issue here is that we want only one substitution to happen each time we go through the loop, so that one _ gets replaced each time. However, our substitution method is too eager. It replaces all of them the first time through so that there are no _ characters left and no other substitutions actually happen.

There is an easy fix for this. The sub() function accepts an optional fourth argument, count, that limits how many substitutions will be made. For instance, if sub(text, 'sky', 'sea', 3) were used, then *sky* would be replaced by *sea* at most three times in the string text. So, what argument do we need to add within the call to sub() in order to fix our function? See if you can figure it out and get this function working.

After you've fixed the function, we need only do one more thing to get a Hopkins cut-up: simply shuffle the words and pass the shuffled list as an argument.

```
from random import shuffle↵

shuffle(things)↵

print(beauty(things))↵

shuffle(things)↵

print(beauty(things))↵
```

I intended to have those last two lines repeated. This will let you see two different shufflings of the catalog.

I must admit that it is very, very obvious to a careful reader that the words in Hopkins's catalog are not arranged at random. Antonyms are placed before and after one another, there are alliterative words, and the words are arranged with respect for how many syllables they have and for the lexical stress. These are some of the most obvious ways in which the poem (or at least, the part of it we are looking at) has been very consciously designed and composed. You can of course just read these lines in the usual way and figure out a lot about why the words are arranged as they are. So why put these words into a blender?

The cut-up method provides another way to experimentally, practically see what happens when the words are exchanged. You can see what you think is maintained about the poem—is it anything?—and what changes. In this case, the changes are many and significant, but are there aspects of sense that remain the same? You'll have to decide for yourself. This is a very miniature, toy example. You can try something more interesting and extensive.

[Free Project 15-2]   Automate Your Own Cut-Up

Do this project at least two (2) times. Find a catalog text of some sort where the words, phrases, lines, or other textual elements have been intentionally arranged for some effect, such as a poetic or rhetorical one. Use an existing text for this project—not one that you yourself have written. Some examples you may wish to consider, to use or to help you identify other catalog writing:

- Christopher Smart's lines about his cat Jeoffry, from *Jubilate Agno*, found among other places at www.poetryfoundation.org/poems/45173
- "Who We Be" by rapper DMX, found in many places on the Web, including at genius.com/Dmx-who-we-be-lyrics
- "VIA: 48 Dante Variations" by Caroline Bergval, which is intentionally organized, although alphabetically; carolinebergvall.com/wp-content/uploads/2018/08/VIA.pdf
- "Girl" by Jamaica Kincaid, listing advice and instructions from a mother to her daughter; newyorker.com/magazine/1978/06/26/girl
- "Matthew XXV:30" by Jorge Luis Borges, which begins (in one English translation), "Stars, bread, libraries of East and West"; search for it online

After you have selected a text, take just the part of it that is "catalogic" and actually is listing things. Segment this text into pieces. This is another conservative cut-up project, in which you will shuffle bits of text that have been cleanly divided from each other. Place these pieces in a Python list. Use a similar technique as with the Hopkins lines, and shuffle the elements, checking a few different outcomes. Are there ways in which the shuffled versions maintain the poetics or rhetoric of the original version? Does the shuffling help to point out why the original order is important, or if it isn't?

[15.9]   Simple Grammars for Text Generation

While we're interested in natural language here, the type of grammar we'll be using in this section is not a linguistic one, but a formal grammar. I will describe some ways to clearly write (or type) such grammars as strings. But for now, let's define what a *formal grammar* is: it's a set of production rules that describe how to produce strings and, specifically, all strings and only those strings that are valid in the syntax of a particular formal language. I'll use *production* (more of a formal language term) and *generation* (more of a natural language generation term) interchangeably in this section because they are directly connected in the way we're using formal grammars.

You already have experience with formal languages because regular expressions define formal languages of a certain type.

You also of course know how to use functions. Although it may not be the simplest way to do it, functions, our familiar old friends, can be used to define a grammar. For instance, consider a very simple language that only contains two strings: the words *right* and *wrong*—nothing else. Those are our only two *sentences* in the language. This function defines the grammar for this super-simple language:

```
from random import choice

def sentence():↵

    return choice(['right', 'wrong'])
```

Perhaps not the most thrilling language, but type this in and try out the `sentence()` function to make sure you understand this very simple example.

Let's elaborate this slightly. We want to also allow two more strings in our language, *rong* because some people may find it funny to have a misspelled word here and *way wrong*; things can't be way right, but they can be either wrong or way wrong. We'll define this grammar so that structurally there is one option, at the top level, corresponding to *right*, and one that branches off to the other three options:

```
def sentence():↵

    return choice(['right', wrong()])

def wrong():↵

    return choice(['rong', 'way wrong', 'wrong'])
```

I'm not claiming this is the ideal way to present this grammar, but it seems like one reasonably clear way to do it. Try this one out as well and confirm that it will generate all four strings in the newly expanded language.

Here is one fairly simple way to type this grammar into a text file, rather than encoding it in these two functions:

S -> 'right' | Wrong

Wrong -> 'rong' | 'way wrong' | 'wrong'

The first production rule is the starting point. (There is a standard convention by which *S* actually stands for *start* rather than *sentence*.) This rule says we can do one of two things; the | symbol is an "or" that means one of several branches can be taken. We could proceed to the *terminal* token 'right': as the name suggests, that means we are done with production; this particular rule has reached a final string. It is indeed a string, with those feet marks (straight single quotation marks) that indicate a string when used in Python. Or we could proceed to the *nonterminal* Wrong. In that case, to continue with the generation of our string we (or the computer) will look through the other rules and find one that applies. It's not very hard to do in this case. The only other rule is the one that applies. That rule says that *Wrong* produces one of these three strings. They are all terminal tokens, so generation of text ceases after this point. Note that there is no sort of formal problem with one nonterminal being called *Wrong* and one of our terminals being *'wrong'*; we might risk confusing ourselves or someone looking at the grammar, but we won't confuse the computer.

With that very simple example out of the way, let's consider something that is still very simple but features a twist. I'll show you a similar sort of written-out grammar representation first, and then we'll consider how to encode the grammar as a set of functions:

S -> Sentence

Sentence -> "we" Aux Act Last

Aux -> "can" | "don't" | "must" | "will" | "won't"

Act -> "find" | "know" | "learn" | "read" | "say" | "see"

Last -> "it" | Sentence

Does this make sense? The first rule has only one possible production, the nonterminal Sentence. (We could rewrite things to omit this rule, but as we will see in a bit, this allows us to have *S* retain its special status as a start token.) The rule for Sentence says that it consists of the terminal *"we"* followed by three nonterminals: *Aux*, *Act*, and *Last*. There is no option, no or in this rule. The generation of text always proceeds by producing that string and following the three rules corresponding to *Aux*, *Act*, and *Last*. *Aux* and *Act* seem fairly similar; they are rules that produce one of a limited number of strings. Every option is a terminal token. But *Last* is different: this rule could result in the terminal *"it"*—but it could also continue generation by moving along to the

Sentence rule. That does something that you may find interesting: It produces another, embedded sentence.

These are some of the strings that are in this language—that is, ones that are generated by this grammar:

we won't say we will learn it

we must say we must know we can say we won't learn we won't learn we don't learn it

we can find it

we will find we will find we can find we can say it

we don't see we don't know it

we won't find it

we must find we don't find we can find it

we won't find we must read we will read we can know it

we will see we won't learn we don't learn it

we won't learn we don't find it

This is a *recursive* grammar. It uses recursion (covered first in 7.6, The Factorial and again in 8.10, Verifying Palindromes with Iteration and Recursion) because sentences can be part of sentences. And this isn't just a complete absurdity: sentences can indeed be part of sentences in English and in other natural languages.

Here's a serious question you should think about and answer: What is the maximum length of the sentences produced by this grammar? To answer it, you might think about this question: In this formal language, what limits the length of a sentence? What rule or set of rules? Or you could ask yourself, is there any sentence that could not have been expanded further if a different choice had been made in generation?

I'm not going to offer a glib answer, or any answer, to this question in this book. Instead, I'll move along to showing how this grammar can be encoded in functions:

```python
def sentence():
    return "we " + aux() + act() + last()
def aux():
    return choice(["can ", "don't ", "must ", "will ", "won't "])
def act():
    return choice(["find ", "know ", "learn ", "read ", "say ", "see "])
def last():
    phrase = choice(["it", sentence()])↵
```

Get all of this typed in correctly and you will have defined a grammar. When you then invoke the following line of code, what you are doing is basically using that first production rule, S –> Sentence:

```
print(sentence())
```

When you do this, you should notice a problem. The computation keeps on going. You can press the Stop button in Jupyter Notebook (to the right of the Run button) to interrupt this process, or wait for the informative error message that will eventually be produced.

This problem arises because in this line:

```
phrase = choice(["it", sentence()])
```

Python calls sentence() *before* it makes the choice between these two elements. Then this call to sentence() results in another call to sentence(), and so on, and so on . . .

There are a few ways to address this problem. I'm going to present the one I think is fairly simple and clear, based on what has been covered in the book so far. I will rewrite the last() function so that at the beginning it picks between "it" or the null string, and only if the null string is selected will sentence() be called:

```
def last():
    phrase = choice(["it", ""])
    if not phrase:
        phrase = sentence()
    return phrase
```

Now see what this grammar generates by calling print(sentence()) many times. Without giving away the answer to my earlier question, previous code lacked a base case. If you need to, return to the earlier discussions of recursion to see what implications that has. There is a 50 percent probability that the function will call itself each time, and that has certain implications, too.

By the way, this example is a Python version of a short, type-in computational poem of mine that appeared first as a tiny Web page, with JavaScript, in *Increment* magazine, increment.com/open-source/wont-you/. Now that you are near the end of the book, you may wish to compare how things are done in JavaScript—the language used at the very beginning of our involvement with code, in chapter 3, "Modifying a Program."

A final note about generating text from grammars: There are existing systems you may wish to use when you program grammar-based generators. The Natural Language

Toolkit has facilities that allow for generating language starting with a grammar in string form, such as:

```
grammar_string = """↵
    S -> 'right' | Wrong↵
    Wrong -> 'rong' | 'way wrong' | 'wrong'"""
```

After installing any NLTK resources you need, feel free to avail yourself of the parse.generate module and the CFG object. More information on this can be found at nltk.org/howto/generate.html. You can also use Kate Compton's Tracery, a JavaScript library for grammar-based text generation found at tracery.io. Tracery has been ported to Python by Allison Parrish, so you can use the Python version if you like, too. Because these are bonus options for generation with grammars, I'll direct you to search online for more information if you wish to pursue these particular methods.

[Free Project 15-3]   An Advanced Text Generator
Explore at least two options in significant depth, completing one project to your satisfaction. Program a creative text generator using some sort of grammar-based technique. You can use the cut-up technique as well, if you like, but in the previous free project you were already asked to do a cut-up text generator, so for this project, integrate your cut-up method with grammar-based generation if you go this route. In case they are useful, here are two examples of fairly simple text generators that use grammars and that are available in Python:

    nickm.com/autopia/

    nickm.com/memslam/random_sentences.html

You should study some of this code, and modify it, to explore how grammar-based generators work. You may use NLTK or Tracery if you like, but you can also write a self-contained generator like one of these two.

[15.10]   Essential Concepts

[Concept 15-1]   Even Words and Sentences Need to Be Defined
A simple concept such as "word count" can be defined in a variety of ways. A reasonable way to determine words, sentences, and the parts of speech corresponding to particular words is using a library such as TextBlob, which is built on linguistic principles and trained on corpora developed by computational linguists. Understand that even state-of-the-art systems to segment text into sentences and tag parts of speech will not

work well on every text, however. Understand, too, how to access these linguistic tools by incorporating TextBlob into your programs.

[Concept 15-2]   Lexical Resources Have Many Uses

From plain-text word lists to elaborate hierarchical systems such as WordNet, there are a variety of lexical resources that can be accessed by programs. Some of the simplest uses of these involve checking on an inflected word list to see if something is a dictionary word and thus correctly spelled. With a more sophisticated resource, such as WordNet, it's possible to do deeper analysis and to generate rich, expressive language that relates to a particular sense of a word. Be sure you understand how to have your program traverse WordNet's hierarchy and move through its senses as easily as you can do so manually in the Web interface.

[Concept 15-3]   More Elaborate Rules Generate Interesting Texts

Beyond the very basic combinatorial generators explored early in this book, it is possible to transform a text (using a cut-up-inspired method, for instance) and generate texts using a grammar. These are still fairly simple, understandable, and rule-based ways of generating texts, but they offer many possibilities. Understand how to take advantage of these methods. Specifically, develop an understanding of how a grammar can recursively generate sentences, just as people can when putting language together in speech and writing.

[16]   Sound, Bytes, and Bits

[16.1]

In this chapter, I will discuss, briefly and at a very low level, how computers produce sounds and how low-level sound code can be written. This is an unusual discussion of sound because it does not involve the higher-level ways the computer can produce notes within a musical framework—by synthesis or sampling, for instance. Instead, I consider how a stream of bytes, directed to an audio device, displaces a speaker to cause those changes in air pressure that we hear as sound. While this will not leave a new programmer with any sophisticated musical capabilities, it can help to very directly ground an understanding of how computers produce sound. At the same time, because there is a correspondence between continually output bytes and the sound waveform, this discussion will connect the inner workings of the computer, at the byte and bit level, with something we can perceive as listeners.

This is a short chapter, not because sound is simple or uninteresting but because it is meant to introduce a refreshing new type of exploration at the end of this book and, for classroom learners, at the end of a semester. You can explore this type of sound and music quickly, and enjoyably, even if you are in a course in which you may be working on a final project.

[16.2]   Introducing Bytebeat

To make this connection, I will focus on one very unusual, very esoteric way of producing sound—and actually music as well—with a computer program: *bytebeat* sound and music. I expect that very few readers of this book will actually go on to develop a musical practice that employs this method. However, it's really pretty easy to start experimenting in this form, whether or not one has musical training.

In fact, while a few fundamentals of sound and computing are explained, one of the fun and important things about this chapter is that it will offer you a chance to experiment with bytebeat expressions in a way that doesn't initially require understanding—but can lead to understanding. This is not unlike the invitation to modify JavaScript, even if is not initially intelligible, in chapter 3, "Modifying a Program."

The essential rules and the technical setup for producing bytebeat music are as follows:

- There is an audio device that accepts a sequence of bytes and produces sound, the bytes defining the waveform that is produced.

- A program is written in which the value in the variable t is ever-increasing—like the time.

- A function in this program accepts only t as an argument and produces a new byte each time t is incremented.

- This function produces the sequence of bytes that is piped to the audio device, causing displacement of the speaker corresponding to each byte.

Practically, this means that you write something like (in Python) `music(t)`, you have an outer loop that starts `t` at 0 and keeps incrementing it, and the output of all of this goes to something like what used to be called `/dev/audio`, an 8 KHz, 8-bit mono-audio channel.

That is, a single channel (it's monaural) that accepts bytes (it's 8-bit) and, specifically, accepts about eight thousand of them each second (it's 8 KHz, which means 8K per second). That 8K, to be exact, is 8192.

At the core, all we need to do is write `music(t)` such that this function (each time it is called) returns a byte.

Although bytebeat is quite low-level and related to the specifics of computers and audio devices, you can also write bytebeat functions that will run in JavaScript and on Web pages and can thus easily share them across platforms. Instead of going through how to write and compile C programs to produce bytebeat music and how to emulate an audio device of the appropriate sort and pipe the output to the device, I'll invite you to use one of these pages and to focus on the core code.

Here are bytebeat pages you can use:

greggman.com/downloads/examples/html5bytebeat/html5bytebeat.html

wurstcaptures.untergrund.net/music/

Because it is explicitly licensed as free software, I am allowed to make the first one of these available on my website, and I've done that at nickm.com/bytebeat_by_greggman/.

Gregg Tavares's HTML5 Bytebeat will be available there in the unlikely event that one or both of the above pages is unreachable at some point.

[16.3]   Bytebeat from Zero

Let's begin by trying to generate some silence, to have the computer produce no sound, even though it may already be doing that. Specifically, let's just output 0 all the time. Here is our expression that always outputs 0:

```
0
```

Play that and listen carefully, especially at the beginning and when you pause or interrupt the program.

Now, try this one:

```
255
```

Again, listen carefully. Afterward, try:

```
127
```

Now, all three of these are generating very similar waveforms, and of course very uninteresting ones. They are keeping the voltage levels constant and the cone of your speaker at a constant position. But if you listen closely and have the volume far enough up, you'll hear clicks when 0 starts playing and stops playing and when 255 starts playing and stops playing. However, I don't hear clicks at the beginning and end of 127.

This is because the cone of the speaker doesn't need to move from its rest position to "output" a constant stream of 127s. But it needs to pop inward to get to the position corresponding to 0 and pop outward to get to the position corresponding to 255 and needs to return to the 127 position afterward. Those movements each make a sound. The bytes you are computing end up actually moving the speaker, to cause vibrations in the air that can be heard as sound. Pretty neat, isn't it? Of course, it's not tremendously different from bytes changing the intensity of a pixel, the light from which reaches your eye.

Let's do something slightly more complex, but still rather simple. We'd like for our bytes to start at 0 and increase to 255, then go back to 0 and increase to 255, and so on, such that the value increases by 1 each tick of time. There's an easy way to accomplish this; we'll just use the following function:

```
t
```

Now, t will grow in size, but it's always being realized as a byte. So when t reaches 256, the byte that is generated will be 0, the next byte realized will be 1, and so on.

If you think about it (or if you try out HTML5 Bytebeat and examine the red line, representing the waveform), you should be able to grasp that this is a *sawtooth* wave. It produces a tone of a certain sort, not the most crisp one.

How could we raise the pitch of this tone? Here's one way:

```
t * 2
```

As you can see visually in HTML5 Bytebeat, the sawtooth wave now has twice the slope that it did before. The wave is at a steeper slope, pitched more steeply, and the sound, too, is at a higher pitch. Try `t * 3` and `t * 4` to see (and hear) what happens.

There's a question you can think about, answer, and then check, using a bytebeat player. But first, a joke:

> Two men, both of them mathematicians, are talking with one another at a restaurant. The first claims that ordinary people do not know even the rudiments of calculus. "Here, I'll show you," he says. "Let's ask the server what the integral of e^x is." He summons the server and poses this problem. Meanwhile, the other mathematician has been quickly jotting "IT'S e TO THE X" on a napkin. He now carefully holds the napkin up, out of sight of the first mathematician. "Hmm . . . ," the server says, apparently deep in thought. "I think the answer is e to the X." The first mathematician appears quite flustered. The second mathematician, smiling, tells her that she's right. She then nods and adds, "plus a constant."

You don't need to remember calculus as well as this restaurant employee, or at all, in order to add a constant in bytebeat and see what happens. What will happen if you add a constant to one of these functions of t? Say, converting the `t * 2` function into `(t * 2) + 1000`? How will the waveform and the sound differ in this case?

[16.4] Exploring Bytebeat, Bit by Bit

Now, since we've multiplied `t` in order to increase the pitch, shouldn't we divide in order to decrease it? We certainly can. Try it out:

```
t / 2
```

```
t / 4
```

```
t / 8
```

At a low level (which is where most bytebeat musicians are focused), there is a way to divide that is more suitable to byte manipulations and is computationally much more efficient. Division is an extremely costly arithmetic operation, taking many processor cycles. Most programmers don't worry about that very much—and probably shouldn't in most cases—but those doing intensive audiovisual programming do. It is faster and more obvious to such programmers to use bitwise operators to accomplish the same thing. So instead of dividing t by 2, an equivalent operation is shifting the *bits* of t right by one bit. This *right shift* is the same thing as moving the decimal point, as discussed in the tax example in 6.2, A Function Abstracts a Bunch of Code, except this time it's a binary point instead of a decimal point. Listen to this right shift:

```
t >> 1
```

I'll describe what is happening, specifically, when t has the decimal value 140. To do so it's important to understand some essential aspects of binary numbers.

While decimal numbers have a ones place, a tens place, a hundreds place, and so on, binary or base 2 numbers have a ones place, a twos place, a fours place, and so on. So you can look at which digits are 1 (which of them are "flipped on") and count up the value of the number: 1100 in binary is, in our usual decimal system, 8 + 4 + 0 + 0 = 12, because the eights place and the fours place are "on" while the twos place and the ones place is "off."

So when t has the decimal value 140, the binary representation of t is as follows:

```
10001100
```

That's 128 (the hundred-and-twenty-eights place is "on") plus 1100, the number just discussed. Shifting that over to the right by one gives:

```
01000110
```

Which is the binary representation of the decimal number 70. Each of the individual values drops down by one place. That leading 1 represented 128 in the original number; it represents 64 in the new number. So every digit except the rightmost one moves down, and (notice that 70 is half of 140) the number is divided in half. Whether that rightmost digit is a 1 or a 0, it's gone after this right-shift process.

Similarly, t >> 2 is division by 4 and t >> 3 is division by 8. If you want to see how the >> operator works, try it out in Jupyter Notebook. >> is implemented as an operator in Python, too. Check 256 >> 1 and 100 >> 1, for instance. If you want to hear how different right-shifts sound, try them in a Web bytebeat player.

You may have guessed that the *left shift* operator, <<, also exists and does the equivalent of multiplication. Give that a try, too.

Another nice bitwise operator is &. Before I describe what it does, just listen to it for a moment:

```
t & 32
```

```
t & 64
```

```
t & 140
```

```
t & 255
```

You should recognize one of these from before because one of them just does the same thing as t. Which one? You should also notice that (among other things) the height of the waveform is capped by the number on the right. For instance, putting & 64 to the right of t means the maximum value will be 64.

The operator & (bitwise and) sets a bit to 1 if *both* the operand on the left and the operand on the right have a 1 in that position. So, t & 127 will effectively cut off the highest-order bit from t, leaving everything else the same. As you should see, there are other effects besides limiting the height of the waveform. Try out some different values and listen to what those effects are.

I'll invite you next to try this expression:

```
t * t
```

And then "slow down" or decrease the pitch of one of those two operands:

```
t * (t >> 3)
```

You should notice that you can use terms to do things similar to the conditional statement. If you use (t & 128), this will be 1 half the time (when the high bit is set, or on) and 0 the other half the time. What if you multiply by this term?

As a final particularly interesting example, I'll describe one of the most famous, independently discovered bytebeat expressions and sketch out how it works. This is the "42 melody" that is the default bytebeat expression in Gregg Tavares's HTML5 Bytebeat player:

```
((t >> 10) & 42) * t
```

Instead of trying to explain it, I'll discuss how you can explore it. We know from the previous two short expressions that t * (t >> 3) is a slowed-down version of t * t. If we wanted to slow this down even further, we could shift the bits right by 10 instead of 3: t * (t >> 10). Listen to that one, and then note that this is just the famous 42 melody without the 42 part. Or, to be more specific, it lacks the & 42 part. So what is that part of the expression doing?

To understand how it adds so much complexity to the sound that is being produced, it's very useful to take a look at the bits that make up this number, at what 42 is in binary. You should find a programming calculator on the Web, or install or locate one on your system, to convert easily between decimal and binary for the next phase of the investigation. You can, of course, convert the number by doing the math yourself, but it's practical to use a computer for this purpose.

In binary, the decimal number 42 is 101010. Let's take just the two least significant bits, 10, which in decimal are represented as 2. Now, listen to what a modified version of the 42 melody, the 2 melody, sounds like:

```
((t >> 10) & 2) * t
```

It sounds like part of what is going on in the 42 melody, doesn't it? Now check the next bit that is "on," the binary number corresponding to 1000. That number, in decimal, is 8. Try & 8 instead of & 2 and listen to that one. Finally, convert 100000 to decimal to get 32 and try out & 32. You can also check your math at this point: 10 + 1000 + 100000 binary is 101010, while 2 + 8 + 32 decimal is 42. You can also listen to what it sounds like when you use the values 1010, 101000, and 100010, which are combinations of two out of the three "on" bits, although I'll leave it to you to convert those numbers to decimal.

These brief discussions will hopefully be enough to invite your experimentation and to show how bytebeat music can be a window into the lower-level workings of the computer.

You should also go and find bytebeat expressions online, where they are available in different contexts (sometimes included in explanatory videos). They are worth trying out, and the ones that interest you are also worth modifying. You can remove parts of expressions, change values, and determine how parts of each bytebeat expression interact with others to produce sounds.

For a more extensive discussion by the inventor/discoverer of bytebeat music, you can read a paper (Heikkilä 2011); there are also a variety of good resources and discussions online.

[Free Project 16-1]   Bytebeat Songs

Do this project at least four (4) times. Develop a bytebeat expression that you find pleasing for some reason. It does not even have to generate "music" in any conventional sense. Try to explain what about the resulting sound you enjoy. Does it have a beat that you like? High-level variation as time progresses? A melody? Harmony? Or perhaps you mainly like the visual representation that appears and can just tolerate the sound.

After you have developed one expression, develop another that does something else, perhaps intensifying what you enjoyed in the first piece or perhaps involving exploration of a different direction.

Beyond what you sought to do differently, do the sound pieces you developed have significant things in common? Could you combine expressions and have them both used to generate sound at once? How? Try it; figure out some means of combining your "songs."

[16.5]   Further Exploration of Sound

Bytebeat music is something of an anomaly in computer music, although a fun one. It also sheds light on how computers work. Those who wish to continue exploring sound can use one of several cross-platform systems for sound synthesis. Pure Data, which is by Miller Puckette and was originally released in 1996, is a patch-based programming language in which objects are visually linked together. Pure Data supports the development of multimedia projects and is widely used. SuperCollider, originally released by James McCartney in the same year, is one of these that is used extensively, including for livecoding (programming in performance). Many other systems that are widely used for livecoding music. FoxDot is one that uses SuperCollider for its sound synthesis but allows performers to livecode in Python.

[16.6]   Essential Concepts

[Concept 16-1]   Arithmetic Can Be Done Bitwise
This work with sound offers a chance to understand operations on bytes and bits as distinct from the usual decimal arithmetic. Whether you end up having recourse to these often, you should develop an improved understanding of how media are represented digitally.

[Concept 16-2]   A Stream of Bytes Is a Waveform
The digital representation of sound in bytebeat music doesn't take a form that relates directly to standard musical notation, but it does relate directly to the material sound system, the way the speaker moves. Understand this connection and be able to determine, for instance, how large of a file (of data—that is, bytes) you would need to produce five minutes of 8-bit, 8 KHz audio that is not generated or compressed.

[Concept 16-3]   Moving a Speaker Produces Sound
When a digital waveform of this sort is converted into magnetic force and used to move the cone of a speaker, the bytes effect those waves in the air that can reach people as sound.

[17] Onward

[17.1]

It's about time to take the training wheels off—but where are they? Even if we did proceed incrementally, covering fundamentals and starting from simple types of computation, I would submit that we have been doing real programming throughout.

You've been using Python, just like professional programmers do. It wasn't running in any sort of kiddie mode. You used Jupyter Notebook and also figured out how to write a Python program in a text file and run it from that file.

You worked in Processing, which was created with programming education, among other things, in mind. But Processing is used to create art, TV commercials, and advanced visualizations for researchers. Processing has a simple but effective IDE, and by using it, you've gained experience programming not only in an interpreter, not only by editing a text file, but also by using one of these specialized interfaces—an integrated development environment.

When you grabbed an HTML page with JavaScript in it from the Web, opened it in a text editor, modified it, and ran it once more by opening the file in a Web browser, you did something plenty of Web programmers have done, with their own code, collaborators' code, and code on the Web that is offered in example snippets or licensed as free software. You picked this project up later in the book and had the chance to reflect on having some distance from your first experience with it.

Of course, one can learn plenty more about programming. There's more to learn about programming in general, and about a language such as Python in particular, and about tools such as debuggers, test frameworks, version control systems, and static code checkers. And, conceptually, there's more that can be learned about functional programming, concepts like generics, object-oriented programming, and the different sorts of complexity that programs can have in time and (storage) space.

But if you've been through this book, programming on your computer along with the discussion, chapter by chapter—rather than just flipping to the end—you should understand that you have indeed been programming and can continue doing so. The activities in this book aren't pretend programming or almost programming. Anyone who has completed this book, or an abbreviated version of the work if that's what fits into a particular course, has programmed a computer.

Programming in this book was done in several languages, so you, the reader, didn't just "Learn Java in 17 Days" or gain narrow skills in some other particular language. (You should be able to see, though, why it is possible for someone with programming background to learn the essentials of a new language in a few days.) If you programmed along with others in a group and showed your work to them, you no doubt saw how your work can be surprising and meaningful to others who you respect. You programmed some interesting and provocative things, and in ways that can lead you, in the future, to better think and inquire about topics that interest you.

I hope the programming you did helps show you why it matters to have access to code and to have software freedom. Although some JavaScript online is obfuscated either to minimize its size or to intentionally prevent its reuse, you couldn't have modified JavaScript code without being able to find some of it online that is open to you. You couldn't have done the same work without being able (at least in an informal, cultural way, for educational purposes) to do whatever you liked with this code. And every system that we used in this book is free software. All of them can be used for any purpose by anyone, and the workings of these systems are fully available for inspection and modification.

A concrete case of this can be seen in WordNet, an English lexical resource that is included in TextBlob. This resource is licensed for use by anyone—academics, industrial researchers, artists, hobbyist programmers, whoever. This is great news for those of us who speak and write English and who want to write programs that deal with English-language texts. What about everyone else?

Fortunately, there are non-English WordNets. Unfortunately, some of them are not free software. If you're a member of the world's largest language community, a speaker of Mandarin Chinese, you are welcome to browse the Chinese WordNet online but can't download and use it in your programs. In German, there are two WordNet projects, one restricted and one open only for academic use (sorry, artists). In Spanish, there is one WordNet project, focused on Iberian Spanish, and it too is payware. The effect of a free software English-language WordNet and restricted systems in other languages is to (further) advantage English, making it less likely that people will do computational linguistics work and creative computing work in other languages. It particularly suppresses exploratory programming work that isn't done as part of funded research projects that can pay for access to such resources. Is this the world we want?

I would guess that, closer to home, if you weren't allowed to figure out a budget for your household or your business because doing those calculations was restricted, you would be troubled by this. So why, when it comes to being able to think and work as an artist or humanist, would anyone settle for locked-down software—for not having full control over their own work and thinking?

To start making the move to free software, you can start by not restricting yourself to a proprietary language that is out of your control—something easy to do, because the leading programming languages are free software. You can also start moving to applications (for word processing, photo editing, and so on) that are free software and developed by a community seeking mainly to empower themselves and others. Ideally, then, a person who believes in intellectual and software freedom would move to a free software operating system. This is an ideal that is within reach; there are plenty of us who have done it.

Onward to thinking about your further development as a programmer. If you do want to become a professional programmer, or a computer scientist—these are not the same thing—there is certainly much more to learn in either case. It will no doubt be an extensive project, although rewarding and, ultimately, possible.

If you want to bring computation to bear on questions that interest you—as an artist, as a humanistic researcher, or as some other type of thinker with complex and serious questions to pose—you can do it; that goal is also within reach. There will surely be more to learn in this case, too. You should understand that having determined a way to explore and to extend your ability to inquire, you can accomplish this. You will need to figure out how to prepare data and formulate useful questions that serve as a starting point. As your projects become more elaborate, awareness of all those things that professional programmers know about may become more important. You may certainly wish to collaborate in different ways and, of course, have other programmers take a role in your projects, perhaps working with you directly in collaborative coding situations.

At this point, though, you are prepared to sketch, brainstorm, and explore using programming by yourself, when an idea strikes you, before anyone else is even involved. You are ready to approach computation and the questions that matter to you not as a manager unaware of how computing is done, but as a programmer. There is much more to learn, indeed, but the computer should no longer seem a black box. Those who have been through this book have implemented some of the underlying workings of Photoshop; creatively conflated two books; built a text classifier; developed a generated, navigable landscape; and used computing to ask questions that no existing site or premade download could have asked. If you have done this or significant portions of it in a class, you have been able to augment your thought using computing. You can build a model of the world—the literary, artistic, historical, material, culturally situated world—and use it to inquire and create in new ways.

To continue your work as a programmer does not mean turning inward to discover every technical detail of computing while neglecting people, culture, communication, and art.

It can mean working with programming, and continuing to learn more about computing, while turning newly learned computational abilities outward, onto the wide world. It can mean seeing the world in a new way—asking new questions of it, questions that would be difficult or impossible to pose without computation. Questions that can lead to powerful new creations and new insights.

Hello, world.

[Appendix A]   Why Program?

[A.1]

Using computation for inquiry is, unfortunately, not yet widely recognized as important in the humanities. Programming on digital humanities projects is still often considered a finishing stage, not a means of determining what to explore or how to proceed. While readers of this book will usually already wish to learn to program, the virtues of programming are not always clear to everyone.

Programming is not simply an implementation detail for interchangeable labor to provide but is related to the methods humanists and artists use to model the world. In all other fields that use computation significantly, including economics and biology, researchers frequently sketch, explore, and frame the nature of their investigations by writing programs. Understanding computation and having basic skills in programming allows researchers to question, refine, overturn, or further develop existing data representations, computational methods, and theories. I discuss cognitive, cultural, and social rationales for programming and argue that the practice of exploratory programming will be critical to innovation and revolution in the digital humanities—that it should be taught at undergraduate and graduate levels and should be used early on in research projects.

[A.2]   How People Benefit from Learning to Program

The book *Digital_Humanities* lists a variety of technical skills, rooted in text encoding and in information technology project management, that are important to the digital humanities. Although determining the appropriateness of scripting languages is listed as important, being able to program is not mentioned. Similarly, *Debates in the Digital Humanities*, a collection that features a section called "Teaching the Digital

Humanities," has nothing to say about whether programming should be taught. Typically when digital humanities (DH) pedagogy is discussed, the real topic is how to use preconstructed DH systems to deliver education. Yet humanities students can surely be offered the same opportunity that Seymour Papert, using Logo (Papert 1980), and Alan Kay, using Smalltalk (Kay and Goldberg 2003), successfully offered to children. Adults studying the humanities and the arts can also be allowed to learn programming.

The case for programming education would not be as strong if programming were merely instrumental, if it involved nothing more than completing an already established plan. In advocating that humanists and artists should program, I consider exploratory programming, which involves using computation as a way of inquiring about and constructively thinking about important issues.

Those who decide to become new programmers often find the motivation to do so in their encounters with computers and in interacting with others who are using programming to think about interesting problems. They often have concrete and personal reasons for engaging with computing and do not need to consult the sort of argument that I present here. The discussion here might, however, help humanities and arts students better articulate their interest in programming to fellow students and to faculty members.

One humanist who has advocated for programming education—both in his writing and by teaching students to engage with programs in humanities classes—is Matthew Kirschenbaum. He argues: "Computers should not be black boxes but rather understood as engines for creating powerful and persuasive models of the world around us. The world around us (and inside us) is something we in the humanities have been interested in for a very long time. I believe that, increasingly, an appreciation of how complex ideas can be imagined and expressed as a set of formal procedures—rules, models, algorithms—in the virtual space of a computer will be an essential element of a humanities education" (Kirschenbaum 2009).

Kirschenbaum is one of several humanists who were teaching programming to undergraduate and graduate students in different contexts before the first edition of this book was published. I have taught programming to media studies master's students at MIT and media studies master's and undergraduate students at the New School, and have MIT undergraduates doing computational writing projects in The Word Made Digital and in Interactive Narrative (my current and older course Web pages are linked from nickm.com/classes). Daniel C. Howe developed and taught the course Programming for Digital Art and Literature (rednoise.org/pdal/) at Brown and RISD. At Georgia Tech, Ian Bogost has taught courses that include a Special Topics in Game Design and Analysis section (syllabus at bogost.com/teaching/atari_hacks_remakes_and_demake/)

devoted to programming the Atari VCS. Allison Parrish teaches Python programming in Reading and Writing Electronic Text (rwet.decontextualize.com), regularly offered in NYU's Interactive Telecommunications Program (ITP). Others who have taught programming to humanists include Michael Mateas and Stephen Ramsay. There are also many courses for artists and humanists in Processing, which was created by Ben Fry and Casey Reas to help designers learn programming and is ideal for developing interactive sketches.

In the following sections, I offer arguments that programming

- allows us to think in new ways,
- offers us a better understanding of culture and media systems, and
- can help us improve society.

After this, I'll return to the ways that programming can be enjoyable, explaining what special qualities of programming may make it a particularly pleasing way to occupy our time and to contribute new creative work to the world.

[A.3] Cognitively: Programming Helps Us Think

One useful perspective on how computing can improve the way we think has been provided by educational researchers, who chose to "distinguish between two kinds of cognitive effects: Effects *with* technology obtained during intellectual partnership with it, and effects *of* it in terms of the transferable cognitive residue that this partnership leaves behind in the form of better mastery of skills and strategies" (Salomon, Perkins, and Globerson 1991, 2).

The first of these effects is obvious in many domains. The person using a spreadsheet to try out different budgets and scenarios is better prepared to innovate in business than the person who lacks such a system and must calculate by hand. A civil engineer modeling an unusually designed bridge with a computer is better able to ensure that it is safe than is one who must rely on earlier methods. A radiologist using a modern, computational MRI system is able to deliver a diagnosis in cases where X-rays would not be adequate.

This positive effect of computation is what computer pioneer Douglas Engelbart called "augmenting human intellect" (Engelbart 2003). Although there are very many domains in which thinking *with* computers have proven effective, some people are nevertheless resistant to the idea that thinking with computers can be helpful in the humanities and arts. However, computing can be used to model artistic and humanistic processes, just as it can be used to model business and economic processes, bridges from an engineering perspective, the human body from a medical perspective, and so on. Thus, programming has the potential to improve our humanistic and artistic thinking as well.

There is also hope that thinking computationally can enhance the way we think more generally, even when we are not using computers. Indeed, there is evidence that adding computational thinking to the mix of our experiences and methods can improve our general thinking. Perhaps an obsessive focus on programming could be detrimental. But those who have background in the arts and humanities and who choose to learn programming are diversifying their ways of thinking, adding to the methods and perspectives that they already have. Programming can help them consider the questions they care about in new ways.

The research that has been done about whether programming improves cognition has focused on younger learners who are still developing cognitively, not students in higher education. Nevertheless, to provide some insight into the effects of computer programming, I offer some results from the literature on whether learning to program can help people of that age group improve their cognition.

[A.3.1]   Modeling Humanistic and Artistic Processes Is a Way of Thinking

Edward Bellamy, in *Looking Backward: 2000–1887*, projected a character more than a hundred years into the future to explain the author's utopian vision of society. Similarly, Douglas Engelbart wrote about how computation could augment human intellect in a more or less science fictional mode. Writing in the voice of a hypothetical augmented human, Engelbart explained more than fifty years ago why people using computers as tools (even if they were using the advanced technology that he envisioned) should understand computer programming:

> There are, of course, the explicit computer processes which we use, and which our philosophy requires the augmented man to be able to design and build for himself. A number of people, outside our research group here, maintain stoutly that a practical augmentation system should not require the human to have to do any computer programming—they feel that this is too specialized a capability to burden people with. Well, what that means in our eyes, if translated to a home workshop, would be like saying that you can't require the operating human to know how to adjust his tools, or set up jigs, or change drill sizes, and the like. You can see there that these skills are easy to learn in the context of what the human has to learn anyway about using the tools, and that they provide for much greater flexibility in finding convenient ways to use the tools to help shape materials. (Engelbart 2003, 93–94)

Engelbart presents one way of understanding the computer metaphorically, as a workshop that allows people to build things. Not being able to program is akin to not being able to change a drill bit. A person *can* use a workshop in such circumstances but is limited. Another way of understanding the computer is as a laboratory. If people can use the equipment that is there but are unable to change the experimental setup, then they are limited in what experiments they can do. Seeing the computer in these ways, as a means of thinking constructively or experimentally, helps to explain why people who are artists and critical thinkers would want to be able to adjust computation in a variety of ways. Such adjustment was done in Engelbart's time, and still is done, by computer programming.

Perhaps the most problematic aspect of the statement by the hypothetical augmented human is the mention of how people often "learn anyway" about aspects of programming. Environments for programming (typically, BASIC) became easily available to the everyday home computer user in the late 1970s and 1980s, but as powerful, complex IDEs and compiled languages have been developed, programming has in some cases become more difficult to access. People do not encounter it casually in the ways they used to. This means that some unnecessary complexity has been hidden, which is a positive outcome, but it also means that some flexibility has been removed. Still, those who delve into HTML, learn to use regular expressions to search documents, and start to develop short shell scripts do end up gaining some familiarity with their computational tools and can build on that to begin to learn skills relevant to programming.

Engelbart's work focused on improving complex processes and on facilitating teamwork and was also very engaged with building models of salient aspects of the world. While Engelbart was not focused on humanistic and artistic work, constructing computational models is useful in the arts and humanities as much as in economics, biology, architecture, and other fields. One way to frame this sort of model-building in the humanities and arts is as *operationalization*, and this was the term used at the Media Systems workshop in 2012. As the report of the workshop noted, "operationalization almost always involves novel scholarship both in computational systems and in the area being modeled." Unfortunately, "few individuals are prepared to do both types of research, while interdisciplinary teams are difficult to assemble and support" (Wardrip-Fruin and Mateas 2014, 48–49). If exploratory programming were undertaken more often by humanists and artists, these explorers and programmers would be able to do this work of operationalization more easily, both individually and in collaborating teams.

Systems of these sorts, whatever domain they are in, inherently embody arguments about the theories they draw upon. For instance, at the most abstract level, they seek to show what parts of a theory can be formalized and what that formal representation should be. Such systems, by virtue of how they are constructed, also argue that certain aspects of a theory are independent and others are linked. These models can be used for reflection by scholars and researchers, for poetic purposes (to make new, creative works), or for study. However, a computer implementation by itself, even without a "human subjects" experiment, is a way of engaging with a theory and attempting to understand and apply it in a new way.

[A.3.2] Programming Could Improve Our Thinking Generally

Considerable educational research was undertaken in the 1970s and 1980s to assess the value of computers in grade school education; some of this focused on computer programming specifically. The results varied, but in 1991 a meta-analysis of sixty-five of them, which involved coding the results from each and placing them all on a common scale, was published. It considered quantitative studies available in university libraries that took place in classrooms (at any grade level) and assessed the relationship between computing programming and cognitive skills (Liao and Bright 1991, 253–254):

> The results of this meta-analysis indicate that computer programming has slightly positive effects on student cognitive outcomes; 89 percent of positive study-weighted [effect size] values and 72 percent of positive ESs overall confirm the effectiveness of computer-programming instruction. . . . Students are able to acquire some cognitive skills such as reasoning skills, logical thinking and planning skills, and general solving skills through computer programming activities. (Liao and Bright 1991, 257–262)

The researchers noted that the effect was moderate and that their analysis did not assess whether computer programming was better to teach than were other alternatives. Also, the study was assessing grade school education research rather than programming education in colleges and universities. Still, the conclusion was that, at least for young learners, there were observable cognitive benefits to learning programming.

This meta-analysis also determined that the benefits of learning to program could go beyond a specific programming language. However, it suggested that the selection of an appropriate language was important because programming education with Logo had the greatest effect size (Liao and Bright 1991, 262). Logo was not used exclusively for exploratory programming in the 1970s and 1980s, but I suspect that its use was significantly correlated with an exploratory programming approach, which was part of Seymour Papert's original vision for the language. So I read these results as consistent with (although not clearly demonstrating) the value of exploratory programming in particular.

It's true that these are K–12 studies, and the instruction provided was almost certainly either limited to learning about programming itself or to doing math. The effect, too, was not a strong one. The significant gains from programming education—as determined in this analysis of sixty-five studies—are quite relevant to the arts and humanities, however. Older students are developmentally different—and yet what if opportunities remain to improve students' "reasoning skills, logical thinking and planning skills, and general solving skills"? Doesn't that, by itself, speak in favor of teaching programming as a method of inquiry in the humanities and arts? Do any of the other humanistic methods that we teach to these advanced students offer documented, general cognitive benefits, observed at any grade level?

[A.4] Culturally: Programming Gives Insight into Cultural Systems

The argument here is twofold. First, as critics, theorists, scholars, and reviewers, those who have some understanding of programming will gain a better perspective on cultural systems that use computation—as many cultural systems increasingly do. Second, after learning to program, people are better at developing cultural systems as experiments about, interventions into, augmentations of, or alternatives to the ones that already exist.

[A.4.1] Programming Allows Better Analysis of Cultural Systems

Douglas Rushkoff writes: "For the person who understands code, the whole world reveals itself as a series of decisions made by planners and designers for how the rest of us should live" (Rushkoff 2010, 140). By understanding how media and communications systems are programmed, we gain insight into the intentions of designers and the influence of material history, protocols, regulations, and platforms. In many cases, a full understanding of, for instance, a Web application will involve understanding not only the decisions made by the developer of that application, but also the decisions that have been made in creating and upgrading underlying technologies such as HTML, CSS, and programming languages (JavaScript, PHP, Java, Flash).

Consider a few questions related to culture and computing: Why do many games for the venerable Nintendo Entertainment System share certain qualities, while different qualities are seen in even earlier Atari VCS games? How do the options offered for defining video game characters, virtual world avatars, and social network profiles relate to our own concepts of identity? How does word processing software, with its formats, typographical options, and spell- and grammar-checking, relate to recent literary production? How have tools such as Photoshop participated in and influenced our visual culture? How did a small BASIC program exist in cultural and computational contexts and have meaning to computer users of the 1980s? Because the cultural systems relevant to these questions are software machines built out of code and hardware machines made to be programmed, knowledge of programming is crucial to understanding them.

Scholars in the humanities of have already used their knowledge of programming and their understanding of computation to better understand the history of digital media. Extensive discussion of this sort has been provided in book-length studies. These include studies of early video games by Nathan Altice (2015) and Nick Montfort and Ian Bogost (2009); of identity in digital media by D. Fox Harrell (2013); of word processing by Matthew G. Kirschenbaum (2016); of Photoshop and visual culture by Lev Manovich (2013); and of a one-line Commodore 64 BASIC program by Montfort et al. (2013). In several of these cases, the methods of inquiry these scholars used included developing software and learning from the process of programming. In all of these cases, these scholars brought their understanding of computing—developed in part by doing at least some amount of programming and exploration—to bear on these questions. While these particular studies have been done, many open questions remain regarding how these and other programmed systems participate in our culture.

[A.4.2]   Programming Enables the Development of Cultural Systems

To ground this aspect of programming in practical concerns, consider that by learning to program, people enlarge their ability to develop new cultural systems and to collaborate on their development. Michael Mateas, writing of his experience developing a course in programming (one aspect of procedural literacy) for artists and humanists, explains how an awareness of computation allows work on new sorts of projects:

> Procedurally illiterate new media practitioners are confined to producing those interactive systems that happen to be easy to produce within existing authoring tools. . . . collaborative teams of artists, designers and programmers . . . are often doomed to failure because of the inability to communicate across the cultural divide between the artists and programmers. Only practitioners who combine procedural literacy with a conceptual and historical grounding in art and design can bridge this gap and enable true collaboration. (Mateas 2005)

Mateas is not simply claiming that artists and humanists should learn computing jargon so as to be able to bark commands at programmers. He is discussing communication at a more profound and productive level, the sort that allows for the exploration and expression of new ideas.

To close the "two cultures" gap that Mateas identifies in new media and the digital humanities, it would of course be ideal for those who are technically expert to learn some about the humanities as well. While the methods and goals of humanistic research may not be obvious to all programmers, it is quite difficult to find programmers (at least, ones in the United States) who have never taken a course in the humanities at all, who have never studied a novel or taken a history course. It is still easy, however, to find artists and humanists who have no experience with programming.

[A.5]   Socially: Computation Can Help to Build a Better World

Programming cannot only contribute to social and utopian thought; I believe it is also uniquely suited to building productive utopias. I consider a *utopia* to be a society (usually represented or simulated in some way, although there are utopian communities that are actual societies) that is radically different from our own and yet is also engaged with our own society. A utopia might be an attempt to provoke people, or it might be offered as a serious model that could be emulated. In any case, a utopia is not an escapist vision, nor it is it an alternative place with no relation to our society, the sort of place that has been called an *atopia*.

Utopias don't have to be perfect to be useful to social and political thought. In terms of provoking people to think about important issues in new ways, utopias can be presented that are worse than our current society. These are called *dystopias*; because they present arguments about how our society might improve, I consider them to be in the broad category of utopias as well.

Programming can be used to develop utopias via computer games and simulations. The original *Sim City*, for instance, can be read as a model city that promotes mass transit and nuclear power. Modified versions of it could present other simulated societies, using computation to make different arguments. Gonzalo Frasca makes the argument that these types of simulation games could, if open to modification, become the "Sims of the Oppressed" along the lines of Agosto Boal's "Theatre of the Oppressed," allowing people to model and discuss social and political aspects of life (Frasca 2004). Programming can also enable new social spaces and developments, such as pseudonymous online support groups that are open to people around the world.

Both types of potential are indicated by Douglass Rushkoff: "We are creating a blueprint together—a design for our collective future. The possibilities for social, economic, practical, artistic, and even spiritual progress are tremendous" (Rushkoff 2010, 14). To take this idea seriously, rather than cynically dismissing it: If we are to be designers of our collective future, what does that sort of design entail, and what skills should we have to participate in this collaborative activity?

Rushkoff offers his answer—that we should fully develop our ability to write online, using computers:

> Computers and networks finally offer us the ability to write. And we do write with them on our websites, blogs, and social networks. But the underlying capability of the computer era is actually programming—which almost none of us knows how to do. We simply use the programs that have been made for us, and enter our text in the appropriate box on the screen. We teach kids how to use software to write, but not how to write software. This means they have access

to the capabilities given to them by others, but not the power to determine the value-creating capabilities of these technologies for themselves. (Rushkoff 2010, 19)

Given this perspective, it seems hard to justify that developing social media wiles specific to whatever the current proprietary systems are—the ability to skillfully use Facebook or Twitter, for example—really constitutes the core skill for the collective designers of our future. It sounds like arguing that we will be able to develop a progressive new society because we know how to navigate our local IKEA. If we envision ourselves as empowered to determine a better future together, we will need to know much more than navigation, more than how to shop, consume, select, and inhabit existing corporate frameworks. We will need to know how to participate in creating them, whether the goal is incremental development or a radical provocation. In Engelbart's terms, we will need the full use of our *home workshop* to have all of the tools available to us and adjustable.

Programming ability has been used to develop new cultural systems, of course. One example is a system launched in 2009 by a for-profit company. This system, Dreamwidth, aimed to correct problems with LiveJournal, which runs on free software code, by forking that code to create a new system. The company improved the way the site could be accessed on screen readers, provided a different privacy model for journal viewing, and published the first widely discussed diversity statement. This cultural system was developed with a focus on writers, artists, and others who were contributing creatively. The community of developers that works on the Dreamwidth code (and, because this is free software, also has full access to this code for any purpose) is remarkable. By the first year after launch, half of the developers were people who had never programmed in Perl or contributed to a free software project before, and about 75 percent were women (Smith and Paolucci 2010). To put this in perspective, as of that year, estimates of the percentage of women participating in free software projects overall ranged from 1.5 percent to 5 percent (Vernon 2010).

The Dreamwidth response involved not just a verbal critique of the problems with LiveJournal (where the Dreamwidth cofounders worked, initially); it also involved more than just producing a proposal or mock-up of what might be better. The response was a project to build a new system with the participation of programmers, including many new programmers. The result was a site that hosts a diverse community and an inclusive group of developers.

With that specific example in mind, consider one more statement from Rushkoff about the importance of participating in and humanizing computing: "The more humans become involved in their design, the more humanely inspired these tools will end up behaving. We are developing technologies and networks that have the potential to reshape our economy, our ecology, and our society more profoundly and intentionally than ever before in our collective history" (Rushkoff 2010, 149).

[A.6]   Programming Is Creative and Fun

At the risk of trivializing what I understand as a cognitively empowering practice, one that is capable of providing us better cultural understanding and one that can help us build a better society, it would be remiss of me not to mention that programming is an activity that gives the programmer poetic pleasure, the pleasure of making and of discovery through making. I discussed earlier how programming is not *only* a hobby to fill the time; this particular aspect of programming that I am discussing now is indeed connected to some types of productive hobbies, as well as to artistic practices. It's worth noting that there are special creative pleasures of programming.

It is enjoyable to write computer programs and to use them to create and discover. This is the pleasure of adding something to the world, of fashioning something from abstract ideas and material code that runs on particular hardware. It involves realizing ideas, making them into functional software machines, in negotiation with computational systems. The strong formulation of this impulse to make and implement in the digital humanities specifically is the declaration that the only true digital humanists are those who build systems ("hack") rather than theorize ("yack"; Ramsay 2011). To note that programming is creative and fun, however, does not require excluding other types of involvement in a field, nor does it mean that it is not also fun to critically or theoretically yack. It simply involves admitting the pleasure and benefits of hacking, of exploring with programming.

Writing a program offers enjoyment that is not entirely unlike other types of making in the arts and humanities: the way sound and sense grow and intertwine on the lattice of a poem; the amazing configuration of voices, bodies, light, and space in a play; the thrill of new connection and realization that can arise from a well-constructed philosophical argument. In this way, the activity of programming can be consistent with more traditional activities: writing, developing arguments, creating works of art.

[Appendix B]   Contexts for Learning

[B.1]

Even though this book contains detailed discussion of programming principles, I've tried to be as flexible as possible when introducing programming. I've explained ways of working with different media forms so that someone more interested in text can continue and pursue that sort of work, while someone who cares more about images can continue in that vein. Those who want to use programming mainly to create art can pursue that direction, while those who want to work analytically can do so—and those who have both sorts of interests can continue to pursue them both, too. Part of that flexibility is allowing free projects that, within some constraints, are student-directed and allow new programmers to establish their own goals.

Another way in which this book is meant to be flexible is that it can be used as a textbook in a formal, semester- or quarter-long course, but can also be used less formally, in shorter workshops, by informal groups meeting online or in person, or by individuals.

When I teach classes and workshops, I like to begin by having people immediately modify JavaScript, to show that fun and interesting results can be produced easily and to get across the point that, materially, programming is no more than editing text. I have found it just fine to even introduce the concept of the course and discuss the syllabus after people do this work and share it with each other. This project is basically chapter 3, "Modifying a Program," which can be done again after an initial encounter with it in class.

Instead of using class time to introduce material that students are then expected to read, I ask students to go through chapters of the book first—using all the methods for learning, including doing the exercises and the free projects. The class meetings can then focus on two aspects of learning: whatever was unclear or too difficult to accomplish, and whatever positive work students did accomplish in their free projects. In classes that are the size of workshops or seminars, those who gained a solid understanding of the material can help those who are less certain. The students who benefit from extra discussion and explanation can return during the next class meeting to share their completed or expanded free projects. Because the material in the book is varied in terms of how difficult each chapter is for different students, what media the chapters focus on, and so on, students who take an extra week to present a free project that satisfies them are not particularly behind in the course. They may find the next chapter fairly easy and end up helping other students understand it through class discussion.

Classes may meet three or two times per week, or just once per week. Personally, I find it good to have a single class session of about three hours because there is a good variety of work to do and this provides a nice block of time for critique of student projects and discussion of concepts. If there is more time available during the week because lab, recitation, or studio hours have been allocated, students can use the time to work on their free projects, and the instructor can answer questions (or better yet, help students answer them using available resources) and offer desk critiques of work in progress.

I conclude the semester not with a typical large-scale term project but with what I call a *final exploration*, in which students *return* to an earlier project and revise, rework, and extend it using what they have learned since then. This should be a substantial revision, oriented toward a further exploration and not just polishing something up. But it shouldn't start from scratch. This helps students revisit their earlier learning, see how far they have come, and think critically about using programming for the intellectual goals of inquiry and creativity.

A minor textual change, but a very important emphasis in learning, is that in this edition I explicitly instruct learners to do the free projects a certain number of times, almost always more than once. A single exploration is not much exploration, and even those who are studying in a classroom and see other student projects will have a difficult time thinking about how their own programming process differs as they pursue different ideas. This emphasis on trying out several different approaches will also help students who feel they have to make a perfect, polished product. For purposes of learning exploratory programming, and discovering how programming connects to one's interests in the arts and humanities, I consider that it's much better to spend the same amount of time creating four radically different clocks (in free project 14-3, Novel Clocks) than it would be to spend the same amount of time making one that seems really good.

What follows is a plan for allocating about 2.5 to 3 hours of weekly class time and for planning the programming students will do outside of class.

[B.2] Semester-Long (Fourteen-Week) Course

(Week 1)   [1] Introduction; [2] Installation and Setup; [3] Modifying a Program

(Week 2)   [4] Calculating and Using Jupyter Notebook; [5] Double, Double

(Week 3)   [6] Programming Fundamentals

(Week 4)   [7] Standard Starting Points

(Week 5)   [8] Text I: Strings and Their Slices

(Week 6)   [9] Text II: Regular Expressions

(Week 7)   [10] Image I: Pixel by Pixel

(Week 8)   [11] Image II: Pixels and Neighbors

(Week 9)   [12] Statistics, Probability, and Visualization

(Week 10)   [13] Classification

(Week 11)   [14] Image III: Visual Design and Interactivity

(Week 12)   [15] Text III: Advanced Text Processing; begin final explorations

(Week 13)   [16] Sound, Bytes, and Bits; work on final explorations

(Week 14)   Final exploration presentations; [17] Onward

In a ten-week quarter, with similar amounts of time devoted to the class each week, less can be done, but an abbreviated version of the course is still possible. This plan is for a faster pace at the beginning; it omits two of the later chapters. Some exercises and sections of chapters 14 and 15 may be omitted to make room for final exploration work and presentations on the last day.

[B.3] Quarter-Long (Ten-Week) Course

(Week 1)   [1] Introduction; [2] Installation and Setup; [3] Modifying a Program; [4] Calculating and Using Jupyter Notebook

(Week 2)   [5] Double, Double; [6] Programming Fundamentals

(Week 3)   [7] Standard Starting Points

(Week 4)   [8] Text I: Strings and Their Slices

(Week 5)   [9] Text II: Regular Expressions

(Week 6)   [10] Image I: Pixel by Pixel

(Week 7)   [11] Image II: Pixels and Neighbors

(Week 8)   [12] Statistics, Probability, and Visualization

(Week 9)   [14] Image III: Visual Design and Interactivity

(Week 10)   [15] Text III: Advanced Text Processing; [17] Onward

It's important in any class to determine if students are understanding the essential concepts in each chapter—whether with "mud cards," quizzes or other evaluation instruments, discussion in class, or simply close examination of submitted student work. If concepts aren't clear for many students, there's no point in progressing. Rather than move on before people are ready to progress, it's better to cover less material but understand the underlying concepts, by removing topics/chapters from the end of the course and going over topics again. It's great to reach the end of the course, to program animation and use interaction in ways that are exciting for new programmers and to get to understanding some of the internals of computing as they align with producing sound. To learn to program, however, students need to fully understand iterating over sequences, developing functions, and the differences between types so that they are ready to work with those aspects of programming in any situation.

[B.4]   One-Day Workshop

In a one-day workshop that lasts the full day, it's possible to at least sample a bit of interesting text or image work, particularly if the material in chapter 7, "Standard Starting Points," is omitted or considered briefly. If one's goal is to help people better critique programs, as critical code studies strives to do, it could be better to devote the last phase of the workshop to close consideration of this chapter and not worry about beginning the text or image work.

[B.5]   Individual and Informal Learning

The book is organized for individuals to go through it in order. For an individual reader, it makes sense to read chapter 1 first, get through chapter 2, "Installation and Setup," and then start on chapters afterward that involve continually working with the computer as a programmer.

The main suggestion I have for an individual learner is to find at least one other person who can work with you as you learn programming. It's extremely useful to discuss the material in the book, to be able to name the constructs and concepts while thinking through programming. It's also great to be able to share one's free projects with others. Even if you aren't part of a group going through the book together, you should see if you can show your projects to others and elicit some responses, comments, thoughts, and suggestions.

[B.6] A Final Suggestion for Everyone

I have taught this material over the years in a wide variety of contexts. I've worked with many types of learners who have different attitudes toward programming and who come from different sorts of arts and humanities backgrounds. My conclusion is that people do not principally have difficulty with the nature of the concepts. That is, people do not typically struggle to learn how to program because programming is a technical or mathematical practice.

The main difficulty, instead, comes when people expect to be informed and to learn only by listening, watching, and taking notes. Some people are under the impression that they can learn about programming simply by paying attention in class, reading a book, and perhaps having a discussion now and then. For this reason, the problems I see are not with the technical or mathematical part of programming, but with the *practice* part.

Programming is a practice, and it takes practice. Like learning how to drive a car or play a musical instrument or participate in a sport, the learner who does not do the activity—who does not undertake the practice—is not going to succeed. On the other hand, I have personally seen dozens of people who began with no programming background at all accomplish several remarkable projects in the space of a semester, using computation in a compelling way—many compelling ways, in fact.

So more than anything else: drive the car, play the instrument, and play the sport by actually programming the computer, often and like you mean it.

[Glossary]

(). Called parentheses in this book, they must not be confused with brackets, *[]*. They are used in many programming languages in several ways—for instance, to group together parts of an arithmetic expression and to contain the zero or more *arguments* accepted by a *function* or *method*. In Python, they are also used to indicate one type of *sequence*, a *tuple*.

[]. Called brackets in this book, they must not be confused with parentheses, *()*. They are used to indicate a *list* in Python and a very similar data type, an *array*, in Processing and JavaScript.

&, |, <<, >>. See *bitwise operations*.

=. In all programming languages discussed in this book, a single equal-sign is used for assignment, to associate a *value* with a *variable*. As unusual as it may seem, the single equal-sign is not an assertion that one things equals another, nor is it a test for equality.

==. In all programming languages discussed in this book, two equal-signs are used to test for equality, to ask whether a *variable* has a particular value or whether one *variable* has the same *value* as another *variable*. The result is a Boolean. In case you see three equal-signs used in JavaScript, that's not a mistake. It is an even stricter equality test, only true if there is the same *type* of data on each side.

abstraction. Representing the essence. In computing, abstraction is often concerned with generality: How can a specific computation, a specific portion of *code*, be developed that works appropriately not only for a particular piece of data, but in many different circumstances? Figuring this out involves determining the essence of this computation, as opposed to the details that happen to apply in the immediate situation.

alpha. One of four "colors," along with red, green, and blue, which represents how opaque or transparent a particular *pixel* is.

argument. A means of passing data to a *function* or *method*. When defining a function or method, and when calling it, arguments are always enclosed in *parentheses*. Some functions accept no arguments, but an empty pair of parentheses is still needed. Within the function or method, arguments are used the same way that *variables* are.

array. A data type very similar, although not identical, to a *list* in Python and other *programming languages*.

assignment. See = near the beginning of this glossary.

attribute. Like a *variable*, but belonging to a *class* or an *object*, an attribute holds data. It could also be said to store or label data, as a variable does. Attributes are how objects store data, just as *methods* are the way objects store computation.

block. In precise use, a unit of *code*, which may be several lines long, that is grouped together, for instance so that a loop proceeds over all of it or a conditional statement causes all of it (or none of it) to be executed. In Python, a code block is indicated with indentation, while in Processing and JavaScript, curly braces, {}, are used.

Boolean. A *type* that can be either true or false. In Python, these are represented as the capitalized values `True` or `False`, while in Processing and JavaScript they happen to be all lowercase, `true` or `false`. Such a value results from checking to see if an equality (such as x == y) or inequality (such as x > y) holds. Unlike other *types*, the name of this one is capitalized, because it is named after a person, George Boole.

bounded. Refers to a loop that is going through each element in a sequence or counting up to some fixed limit. For instance, in the Python loop that has as its top line `for i in [10, 20, 30]:` the loop is bounded, and the *code* within the loop will be called only three times. There are also *unbounded* loops.

bit. A *binary digit*, either 0 or 1, corresponding to the smallest unit of digital information.

bitwise operations. While + (addition) and * (multiplication) are operations that add and multiply decimal numbers, the bitwise operations work on *bytes* in ways that are specific to binary numbers. For instance, << is *left shift*, and by shifting every *bit* over one position to the left, it changes 00100111 to 01001110.

brackets. See *[]* near the beginning of this glossary.

byte. A low-level unit of data corresponding to a binary number—specifically, eight *bits*.

bytebeat. A kind of computer music produced by a function that accepts the time as an argument and outputs *bytes*.

call. To call (or invoke) a function or method is to transfer execution to it, also handing it data if the function or method accepts any arguments. When the function or method finishes running, it will return, so that execution continues right after the call.

casting. Changing the *type* of some data—for instance, by making an integer variable num into a floating-point number with `float(num)` or by making that integer variable into a string with `str(num)`.

class. A blueprint for *objects*. For instance, if particular parcels are represented as objects in a courier service's computer system, the `Parcel` class will describe all the data that can be associated with a parcel in this system, and it will include *methods* for accessing and updating that data. If a recipient phone number is truly required for each parcel in every case, it will make sense to enforce that in the class. A particular parcel object will then be an instance of this class, created through *instantiation*. To understand even the basics of object-oriented programming, including developing one's own classes and instantiating objects using them, will require study beyond this book.

classifier. A system to quantitatively categorize inputs, either into one of two categories (a binary classifier) or into more than two categories. Typically, a classifier is developed using machine learning techniques and trained on data, although some are manually written rule-based classifiers.

code. (n.) Formal text that is processed algorithmically. For our purposes, the term usually refers to a computer *program* or part of such a *program*, although for instance HTML (which specifies the structure of a Web page) and CSS (which specifies the visual appearance of a Web page) are code as well. In this book, we discuss source code, which is human-readable and written by people, although a *compiled* application can be described as consisting of machine code. (v.) To write code.

command line. The text-based environment for computer use, allowing the same sorts of things that are done with the graphical user interface (GUI) to be done by typing commands. On GNU/ Linux and Mac OS, the command line is usually accessed via a terminal; on Windows, the Command Prompt is used. The Anaconda Prompt on Windows is also a command line.

comment. A message added to *code* that is ignored by the *interpreter* or *compiler*, having no effect on how the *program* functions. Comments can be delimited in many different ways in different languages, and sometimes in a few different ways within one language. In Python they can be on lines beginning with # and in Processing on lines that begin //. Comments can be used to annotate a *program*, to help the original author and other human programmers understand why the *code* is written as it is. Comments can also be used to "comment out" lines of *code* that programmers may wish to relocate, revise, or put back into place later.

compiler. A computer *program* that accepts *code* as input and produces an executable file. Strictly speaking, other *programs* such as a linker may be involved in the process, but the main distinction from an *interpreter* is that a runnable *program* results from compilation, whereas an interpreter begins running a program step by step. The Processing IDE provides a way to compile a *program* and run it in a single step.

conditional. The "if" statement, which allows computation to be done only in the case in which some condition, and specifically an *expression* of some sort, is true.

corpus. A "body" of digital documents, digital photographs, or the like.

CSS (cascading style sheet). Used for specifying the visual appearance of a Web page, the information encoded in CSS can be embedded in *HTML* or placed in a separate file.

directory. Equivalent to what appears in a GUI as a folder, this is just a container for files, one that can be placed within other directories. On the *command line*, this term is used instead of "folder."

DRY. "Don't Repeat Yourself." Good programming advice that implores programmers to *refactor* and remove duplicate *code*, which can lead to many sorts of errors as development continues. As you are exploring, it is expedient at times to hastily duplicate code and see if a particular approach will work. Even during exploration, however, it is always a good idea to think through what you have done, as soon as you know what direction you are going to have, and refactor to avoid repetition.

encapsulation. A type of *abstraction* that defines helpful boundaries to prevent unintentional interference between parts of a system.

empty list. The list [ ] that has no elements in it and thus a length of 0. The string that corresponds to the empty list is the *null string*. Usually one reads of *the* empty list because there is only one: the number of ways that zero elements can be included in a list is only one.

equality. See == near the beginning of this glossary.

error. Very generally, a problem with either a runnable computer *program* or a problem that prevents some *code* from being *valid*. In the latter case, a syntax error is one kind of error frequently encountered. This means that the *code* entered does not fit the proper form, as if you were supposed to enter your email address but you typed something that didn't have the @ character in it. An error message provides the best help the *interpreter* or *compiler* can offer, directing the programmer to what seems to be the exact location of the error.

expression. A representation of some sort of *operations* to be performed. For instance, 2 - 4 is an expression with one operation, subtraction, and with the value 4 to be subtracted from the value 2. Expressions can include *variables* (as in `temperature + increase`), can include logical operations (`power_on and cable_connected`), and can have more than one operation in them, in which case *parentheses* are often used for clarification, as in (1 + 3) * 3. Expressions that evaluate to a *Boolean* value are used in *conditionals*. By itself, an expression can be *valid* or not. Even if an expression is valid, it may or may not correctly represent what it is intended to represent. The simplest expressions are just values themselves, such as -2.

file extension. The suffix to a file name, such as .py or .txt, which is one way of indicating the sort of file one is dealing with. File extensions are useful for programmers but are hidden by default on Mac OS X and Windows systems. Note that changing a file extension does not convert between file formats: a JPEG image cannot be made into a PNG image simply by changing .jpg to .png.

floating point. A *type* of number representation that has a decimal point and is of variable precision. Things that can be counted are probably better represented as *integers*, but the mean of several numbers, for instance, is almost certainly better represented as a floating-point number.

for loop. See *loop*.

free software. A concept distinct from "no-cost" software, free software includes several fundamental freedoms, including the freedom to use the software for any purpose and the freedom to inspect, study, and modify it. The term *open source* indicates a closely related concept.

function. A bundle of computation that accepts zero or more values as inputs and returns zero or one values as outputs. In mathematics, a function is a mapping from some domain (e.g., real numbers representing Fahrenheit temperatures) to a codomain (e.g., real numbers representing Celsius temperatures) in which each input has a unique output (which is true in the case of this conversion because no two Fahrenheit temperatures convert to same Celsius temperature). In computing, however, there is no restriction of this sort for functions. A function `capital()` can return `True` for every capital letter and `False` otherwise; the inputs do not need to each map to a unique output. In fact, functions in computing do not need to return a value at all. Functions are usefully thought of as encapsulations of *code* that could have occurred elsewhere in the *program*. They are very similar to *methods*, which are part of *classes* or *objects* and thus can be encapsulated with data.

HTML (Hypertext Markup Language). A formal computer language used to define the structure of Web pages. Web pages written in HTML are a type of *code*, but unless they contain JavaScript, they do not function as *programs*.

IDE (integrated development environment). A special application for writing computer *programs*, allowing a programmer to easily run and debug them.

immutable. Describes a data type whose value (or values, in the case of a sequence) cannot be changed after assignment. The *tuple* is one such data type in Python.

import. The keyword used in Python to indicate a module should be loaded and its components made accessible. Also, conceptually, the process of loading a module.

index. The number corresponding to a particular character position in a string or a particular element in a list or array. Index 0 corresponds to the first character or element, the one that is offset 0 from the beginning.

initialize. To set a *variable* to some initial value. A variable being used to keep count of something, for instance, will usually be initialized to zero.

integer. A number, positive, negative, or zero, that does not have a decimal point followed by digits—for example, 17, -69105, and 0, but not 3.333333. Numbers with decimal points can be represented as *floating-point* numbers.

intentional. Describes a working *program* that does what it is supposed to do at the current stage of its development. Just because a *program* is *valid*, and does something, does not mean that it is intentional. This term, unlike *valid*, is idiosyncratic, but the concept is quite important.

interactive. A *program* is interactive if it accepts user input when it is running.

Internet. The worldwide network of computers connected via the TCP/IP protocol. These days many people write about "the internet," with a lowercase *i*. In popular writing this seems fine and unlikely to cause any confusion. In this book, I've kept the capitalized form because historically, and currently, there are other large-scale networks and internetworks, such as BITNET and FidoNet, and various interbank networks, such as Interac in Canada. From a technical standpoint, however pervasive and ordinary the Internet seems, it is still one particular network just as these others are, and thus a proper noun.

interpreter. A computer *program* that accepts *code* as input and executes it. An interpreter allows one to run source *code*, or a script, directly from a text file. The intermediate step of using a *compiler* is not needed when one uses an interpreter to run *code*. For instance, the python command functions as a Python interpreter (although it actually does some compiling for efficiency), as does Jupyter Notebook. Web browsers interpret JavaScript.

invoke. See *call*.

iterate. To progress through a *loop*, repeatedly executing the *code* in the *loop*. "Iter" means "same," but while the same *code* is executed, the values of *parameters* very often differ. Iteration can be used to count the vowels in a string, for instance, and can be *nested* so as to proceed through all the *pixels* in a two-dimensional image.

lemma. In WordNet, a particular surface form of what we think of as a specific sense of a word. For instance, "bike" and "bicycle" are two lemmas connected to the same underlying sense.

library. A modular collection of code that can be easily reused—for instance, by calling functions and methods provided in it. The term is generally used in many *programming languages*. In Python, a library will take the form of what is technically called a package or module, which can be accessed using an *import* statement.

list. A data *type* that is a mutable *sequence*, very commonly used in Python but also central to several other languages, including Lisp (which stands for list processing). The shortest list, [ ], is the empty list. If a list has elements, and more than one, they are separated by commas between the brackets, as in [34, 28, 23, 14, 8]. Elements of lists can be removed and new elements appended, lists can be concatenated together, and lists can be easily sorted. A list is very similar to an *array*, a data type in Processing and JavaScript.

loop. (n.) Part of a *program* in which *code* is executed multiple times as the *program iterates*. A loop is usually defined with a standard template. It may be *bounded* or *unbounded*. The for loop, with its slightly different syntax in different *programming languages*, is a way to define a bounded loop. An infinite loop is one kind of unbounded loop. (v.) To iterate.

method. A bundle of *code* very similar to a *function* but encapsulated with data as part of a *class* or *object*. An *attribute* is similar to a method, but it holds data rather than *code*.

module. See *library*.

nested. Refers to placing *code* within a control structure such as a *loop*. In particular, a nested loop is a loop placed within another loop. To visit every *pixel* in a two-dimensional image, a *program* will typically use nested loops, for instance to through every possible *y* value and for each such value go through every possible *x* value. For a three-dimensional data structure, another level of nesting can be added.

null string. The string with no characters in it and thus a length of 0. The list that corresponds to the null string is the *empty list*. Usually one reads of "the" null string because there is only one: the number of ways that zero characters can be included in a string is one.

object. A bundle of related computation and data. *Objects* are often used to model particular entities in systems in the world. For instance, a courier service's computer system might model parcels as objects. This would allow the representation of relevant data (such as recipient address and delivery status) and would allow the company to enforce that this data can only be updated in sensible ways (a parcel without a recipient address cannot be given the status of "delivered to recipient," for instance). While we make use of objects and the methods they provide throughout this book, readers will have to do additional study to be introduced to object-oriented programming.

open source. *Programs* whose *code* is available and open to inspection and study. Closely related to the concept of *free software*. Indeed, these terms are sometimes used to mean the same thing.

operator. A means of doing a basic operation, such as the arithmetic operation of addition. In that case, the operator is +. Operators can be used in *expressions* to do computation.

parameter. In computing, a *variable* (or possibly a hard-coded *value* that is manually changed by a programmer during exploration) that can allow an otherwise static or fixed system to vary in certain ways.

parentheses. See *()* at the beginning of this glossary.

part-of-speech (POS) tagger. A computational linguistics system that associates a part-of-speech tag (such as *JJ* for adjective) with each word in the text it is given as input. Taggers today are highly accurate for many kinds of data, but it's important to understand that they do not determine the full syntactical structure of sentences.

pixel. The unit of a bitmapped image, one "picture element" in a grid of them.

polarity. In sentiment analysis, the extent to which some text expresses sentiment at all, for instance, as opposed to being a statement of objective fact.

polymorphism. The ability of *code*, and particularly *operators*, *functions*, and *methods*, to work properly on different data *types*. In Python, + works to add numbers together (including both *integers* and *floating-point* numbers), and it also works to concatenate, or join, *strings*.

program. (n.) A formal specification, written in a *programming language*, of computation to be performed. While a *program*, like other files on disk, is data, it is different from an MP3 or other file in that it is *executable*: because it specifies computation in a formal way, it can run rather than just play or be viewed. (v.) To write programs.

programming. The writing of *programs*, from scratch or by modifying existing *code*. Programming can be exploratory or done to implement a predetermined idea. In all cases but the very simplest, programming is an iterative process in which *code* is tried out, revised, *refactored*, and expanded. The programmer uses modularity, reuse of existing *code*, *libraries*, and other kinds of *abstraction* to accomplish goals, whether these goals are directly instrumental or are ones of art or inquiry.

programming language. A formal system that specifies the syntax in which computer *programs* can be written and the semantics of these *programs* (what they do when executed). Programming languages work on particular platforms (with many of them being cross-platform) and support different approaches to programming.

quantifier. An element of a *regular expression* that indicates how many characters, or grouped expressions, will match.

recursion. A method of computing results that involves decomposing a problem into one part that can be solved easily and a simpler instance of the original problem. A recursive function calls itself to deal with this simpler instance. For instance, to recursively check to see if a list is a palindrome, the part that can be solved easily is checking to see whether the first element and the last element are the same. The part that is a simpler instance of the original problem is checking to see if the "rest" (all but the first and last element) is a palindrome, by applying the same computation to the rest of the list. Eventually, because each instance of the problem is simpler, the recursion will reach a base case and end. For palindrome checking, the base case is a string that is less than two characters long. The typical joke is to offer simply "See *recursion*" as a definition of recursion. However, note that this recursion is broken, as it has no base case.

refactoring. The process of improving the clarity and/or efficiency of *code* without changing the essential computational task that it carries out. By refactoring, the programmer is not trying to make the *code* do something else but to make it do whatever it already does in a way that is more understandable and extensible.

regular expression (regex). A sophisticated means of searching through and matching certain parts of a text that provides a more powerful method than simply searching for a word, but is not as complex as writing a computer *program*.

return. The keyword used to provide a value at the higher level, back to where a *function* or *method* is called, when its execution is complete. Also used to conceptually describe what happens: a function or method returns when it is done.

RGB, RGBA. Two color modes that computers use: (1) red, green, blue; and (2) red, green, blue, *alpha*. Whether there are three or four components, each one has a value between 0 (darkest, or completely transparent for the alpha channel) and 255 (lightest, or completely opaque).

saturation arithmetic. A kind of arithmetic used in computer graphics. When a result is greater than a certain threshold (255 in the case of graphics), it is set as that maximum value. When a result is less than a minimum (0 for graphics) it is set as that minimum. In graphics, this means that an attempt to set a *pixel* as a negative number, representing "super black," will not produce an error; instead, the value will be set to the lowest valid amount, 0.

scope. The limits placed on *code*, defining access to *variables*. For instance, any variables declared within a *function* are local to that function, and changing their *values* will not affect any variables that happen to have the same name outside of that function. Although scope can be difficult for new programmers to understand, it is important because it allows *encapsulation* and modularity, which in turn makes many projects, even exploratory ones, practical.

script. Another name for a *program*, used with those that are *interpreted* rather than *compiled*.

sequence.   An ordered collection of elements, each of which can be accessed by index or by slicing. In Python, *strings*, *lists*, and *tuples* are sequences.

slice.   A portion of a sequence accessed, in Python, by placing [a:b:c] after the sequence, where a is the (optional) starting index, b is the (optional) end index, and c is the (optional) step. Slicing is forgiving, so if there are no characters or elements in the specified slice, a null sequence will be returned and no error will be raised.

source code.   See *code*.

string.   In any modern *programming language*, zero or more characters (such as letters, numerals, punctuation marks) in order. Strings represent textual data. In Python, strings are one type of *sequence*.

syntax error.   See *error* and *valid*.

synset.   In WordNet, a collection of terms that corresponds to one sense of a word, encapsulated in a SynSet *object*. Words that have more than one sense will be part of more than one synset.

tagger.   See *part-of-speech (POS) tagger*.

template.   My term for the very standard and regular structures used in programming languages to define, for instance, *functions* and *loops*. Once a new programmer understands conceptually what function definition is and how loops control the flow of a program, it's possible to do the same basic work in different programming languages simply by understanding how these standard templates differ.

text editor.   An application for creating and editing plain-text files. A word processor is not a text editor and is set up to save documents, by default, in a different format.

token.   A "word" as we think of it when we do a word count. The sentence "The cat is on the mat" has six tokens. Two of them are "the." The other view of a "word" is as a *type*.

tokenizer.   A system for dividing a text into *tokens*.

tuple.   In Python, a data *type* that holds a fixed sequence of elements. A tuple is a sequence, as is a *list*, but it is *immutable*; elements cannot be added to it, removed from it, or have their values changed. Tuples can be indicated with *parentheses* and comma-separated values, such as (1, 2, 3), and are good when one wishes to specify a fixed number of values, such as coordinates in an image and the *RGB* or *RGBA* color values of a *pixel*. *Tuple* is the general term for sequences of any size; those with three elements are called triples, those with two elements pairs, and those with one element singletons. There is also (), the empty tuple. When specifying a singleton, it is necessary to add a comma before the closing parenthesis, like so: (17,).

type.   (n., computing.) The overall category or kind of data that is being represented. For instance, 'hello' is of the type *string*, whereas 17 is of the type *integer*. Types are used to see if certain operations or computations are sensible. For instance, if a *program* attempts to divide 17 by 'hello', this almost certainly results from a mistake on the programmer's part. A type system can help to catch such errors soon after they are made, making development and exploration easier. (n., linguistics) A "word" as we think of it when we look at word usage and figure out which word is most frequent. The sentence "The cat is on the mat" has five types. Of them, "the" is the most frequent as it is used twice. The other view of a "word" is as a *token*.

unbounded.   Refers to a loop that may have a condition for continuing but does not have a fixed number of iterations. For instance, in a Python loop beginning `while True:` the loop is unbounded, and there is no way to determine from the loop itself how many times the *code* under it will be executed. This particular loop seems like an infinite loop, but unbounded loops do not have to take such a form, and here it is possible that some of the *code* within the loop allows for breaking out if some condition obtains—for instance, if the user presses a key or clicks the mouse. In classic, noninteractive punched card computing, it was a mistake to have an infinite loop, which would consume system resources and the paper being used for output. However, creating an infinite loop is not always incorrect today; Web servers, printer spools, and other systems should in many cases run until the process is externally terminated, and plenty of recreational and aesthetic *programs* run infinitely.

valid.   Syntactically or formally correct. For instance, if a person using an automatic teller machine (cash machine) is prompted to enter a four-digit PIN followed by Enter, any four digits followed by Enter constitute a valid (formally correct) response. Pressing the # key or pressing fewer than four digits would be invalid. The question of whether the PIN is the correct one is a separate matter; if it was entered with four digits followed by Enter, then it is of the correct form. *Expressions* by themselves, as well as entire *programs*, can be valid or not. *Programs*, once they are valid, can also be *intentional*, modeling what they were intended to model—but whether they are is a separate question.

value.   A specific piece of data that can be output or used in a computation. For instance, if a *program* should always stop after processing ten items, it might contain `if (count < 10):` in which `count` is a *variable*, while `10` is a value. When this *code* is run, the variable count will (most likely) take on different values each time execution reaches this line.

variable.   A label or slot for a *value*. A variable can "vary" in that it can represent many different values. For instance, the variable count can first have the value 1, then 2, then 3, and so on. Variables can be used in *expressions*. Variables can have any names the programmer chooses; the functioning of the *program* is not affected, although the ability of people to understand it may be.

whitespace.   A general term for tabs, spaces, and newlines, which are used to space out other characters but usually are simply represented as blank areas. *Text editors* and *IDEs* do allow programmers to make whitespace visible, however. In Python, indentation and whitespace is a significant part of the *programming language*, while it does not play the same role in Processing or JavaScript. Those using this book should always use spaces (not tabs) in Python.

[References]

Altice, Nathan. 2015. *I AM ERROR: The Nintendo Entertainment System*. Cambridge, MA: MIT Press.

Burdick, Anne, Johanna Drucker, Peter Lunenfeld, Todd Presner, and Jeffrey Schnapp. 2012. *Digital_Humanities*. Cambridge, MA: MIT Press.

Burroughs, William S. 2003. "The Cut-Up Method of Biron Gysin." In *The New Media Reader*, edited by Noah Wardrip-Fruin and Nick Montfort, 89–91. Cambridge, MA: MIT Press.

Danziger, Michael. 2008. "Information Visualization for the People." SM thesis, Department of Comparative Media Studies, Massachusetts Institute of Technology.

Dobson, James E., and Rena J. Mosteirin. 2019. *Moonbit*. Santa Barbara, CA: Punctum Books.

Engelbart, Douglas. 2003. "From *Augmenting Human Intellect: A Conceptual Framework*." In *The New Media Reader*, edited by Noah Wardrip-Fruin and Nick Montfort, 93–108. Cambridge, MA: MIT Press.

Frasca, Gonzalo. 2004. "Videogames of the Oppressed." In *First Person: New Media as Story, Performance, and Game*, 85–94. Cambridge, MA: MIT Press.

Freedman, David A. 2009. *Statistical Models: Theory and Practice*. 2nd ed. London: Cambridge University Press.

Fry, Ben. 2007. *Visualizing Data: Exploring and Explaining Data with the Processing Environment*. Cambridge, MA: O'Reilly Media, Inc.

Gold, Matthew K., ed. 2012. *Debates in the Digital Humanities*. Minneapolis: University of Minnesota.

Harrell, D. Fox. 2013. *Phantasmal Media: An Approach to Imagination, Computation, and Expression*. Cambridge, MA: MIT Press.

Hartman, Charles O. 1996. *The Virtual Muse: Experiments in Computer Poetry*. Middletown, CT: Wesleyan University Press.

Heikkilä, Ville-Matias. 2011. "Discovering Novel Computer Music Techniques by Exploring the Space of Short Computer Programs." arXiv preprint. arxiv.org/abs/1112.1368.

Hill, Benjamin Mako. 2011. "Freedom for Users, Not for Software." *Benjamin Mako Hill* (blog). October 23, 2011. mako.cc/writing/hill-freedom_for_users.html.

Kafai, Yasmin B., and Quinn Burke. 2014. *Connected Code: Why Children Need to Learn Programming.* Cambridge, MA: MIT Press.

Kay, Alan, and Adele Goldberg. 2003. "Personal Dynamic Media." In *The New Media Reader*, edited by Noah Wardrip-Fruin and Nick Montfort, 391–404. Cambridge, MA: MIT Press.

Kirschenbaum, Matthew G. 2009. "Hello Worlds: Why Humanities Students Should Learn to Program." *Chronicle Review*, January 23, 2009. chronicle.com/article/Hello-Worlds/5476.

Kirschenbaum, Matthew G. 2016. *Track Changes: A Literary History of Word Processing.* Cambridge, MA: Harvard University Press.

Liao, Yuen-Kuang Cliff, and George W. Bright. 1991. "Effects of Computer Programming on Cognitive Outcomes: A Meta-Analysis." *Journal of Educational Computing Research* 7, no. 3: 251–268.

Manovich, Lev. 2013. *Software Takes Command.* London: Bloomsbury.

Mateas, Michael. 2005. "Procedural Literacy: Educating the New Media Practitioner." *On the Horizon—The Strategic Planning Resource for Education Professionals* 13, no. 2: 101–111.

Montfort, Nick, and Ian Bogost. 2009. *Racing the Beam: The Atari Video Computer System.* Cambridge, MA: MIT Press.

Montfort, Nick, Patsy Baudoin, John Bell, Ian Bogost, Jeremy Douglass, Mark C. Marino, Michael Mateas, Casey Reas, Mark Sample, and Noah Vawter. 2013. *10 PRINT CHR$(205.5+RND(1)); : GOTO 10.* Cambridge, MA: MIT Press.

Murray, Janet H. 2011. *Inventing the Medium: Principles of Interaction Design as a Cultural Practice.* Cambridge, MA: MIT Press.

Papert, Seymour. 1980. *Mindstorms: Children, Computers, and Powerful Ideas.* New York: Basic Books, Inc.

Pine, Chris. 2005. *Learn to Program.* 2nd ed. Raleigh, NC: Pragmatic Bookshelf.

Raja, Tasneem. 2014. "We Can Code It! Why Computer Literacy Is Key to Winning the 21st Century." *Mother Jones*, June 16, 2014. motherjones.com/media/2014/06/computer-science-programming-code-diversity-sexism-education.

Ramsay, Stephen. 2011. "Who's In and Who's Out." Paper presented at the History and Future of Digital Humanities panel, Modern Language Association Convention, Los Angeles, CA, January 6–9. stephenramsay.us/text/2011/01/08/whos-in-and-whos-out/.

Reas, Casey, and Ben Fry. 2007. *Processing: A Programming Handbook for Visual Designers and Artists.* Cambridge, MA: MIT Press.

Reas, Casey, and Ben Fry. 2010. *Getting Started with Processing.* Cambridge, MA: O'Reilly Media, Inc.

Richardson, Leonard. 2013. *Alice's Adventures in the Whale*. Crummy, November 18, 2013. Code and text at crummy.com/software/NaNoGenMo-2013/.

Rushkoff, Douglas. 2010. *Program or Be Programmed: Ten Commands for a Digital Age*. New York: OR Books.

Salomon, Gavriel, David N. Perkins, and Tamar Globerson. 1991. "Partners in Cognition: Extending Human Intelligence with Intelligent Technologies." *Educational Researcher* 20, no. 3: 2–9.

Shiffman, Daniel. 2009. *Learning Processing: A Beginner's Guide to Programming Images, Animation, and Interaction*. Burlington, MA: Morgan Kaufmann.

Shiffman, Daniel. 2012. *The Nature of Code: Simulating Natural Systems with Processing*. Edited by Shannon Fry. New York: Daniel Shiffman.

Smith, Mark, and Denise Paolucci. 2010. "Build Your Own Contributors, One Part at a Time." Presentation slides. January 20. slideshare.net/dreamwidth/build-your-own-contributors-one-part-at-a-time.

Tozzi, Christopher. 2017. *For Fun and Profit: A History of the Free and Open Source Software Revolution*. Cambridge, MA: MIT Press.

Vernon, Amy. 2010. "Dreadfully Few Women Are Open Source Developers." *Network World* (blog). March 5, 2010. networkworld.com/community/node/58218.

Wardrip-Fruin, Noah, and Michael Mateas. 2014. "Envisioning the Future of Computational Media: the Final Report of the Media Systems Project." mediasystems.soe.ucsc.edu/sites/default/files/Media%20Systems-Full%20Report.pdf.

[Index]

Page numbers followed by *f* refer to figures.